SpringerBriefs in Applied Sciences and Technology

Computational Intelligence

Series editor

Janusz Kacprzyk, Warsaw, Poland

About this Series

The series "Studies in Computational Intelligence" (SCI) publishes new developments and advances in the various areas of computational intelligence— quickly and with a high quality. The intent is to cover the theory, applications, and design methods of computational intelligence, as embedded in the fields of engineering, computer science, physics and life sciences, as well as the methodologies behind them. The series contains monographs, lecture notes and edited volumes in computational intelligence spanning the areas of neural networks, connectionist systems, genetic algorithms, evolutionary computation, artificial intelligence, cellular automata, self-organizing systems, soft computing, fuzzy systems, and hybrid intelligent systems. Of particular value to both the contributors and the readership are the short publication timeframe and the world-wide distribution, which enable both wide and rapid dissemination of research output.

More information about this series at http://www.springer.com/series/10618

Anthony Mihirana De Silva
Philip H.W. Leong

Grammar-Based Feature Generation for Time-Series Prediction

 Springer

Anthony Mihirana De Silva
Electrical and Information Engineering
University of Sydney
Sydney, NSW
Australia

Philip H.W. Leong
Electrical and Information Engineering
University of Sydney
Sydney, NSW
Australia

ISSN 2191-530X   ISSN 2191-5318   (electronic)
SpringerBriefs in Applied Sciences and Technology
ISBN 978-981-287-410-8   ISBN 978-981-287-411-5   (eBook)
DOI 10.1007/978-981-287-411-5

Library of Congress Control Number: 2015931243

Springer Singapore Heidelberg New York Dordrecht London

Printed on acid-free paper

Springer Science+Business Media Singapore Pte Ltd. is part of Springer Science+Business Media
(www.springer.com)

To Amma and Thaththa

Anthony Mihirana De Silva

To Norma, Priscilla and Nicole

Philip H.W. Leong

Preface

The application of machine learning techniques to time-series prediction is a challenging research problem, particularly for the case of real-world problems which are often nonlinear and non-stationary. An important preprocessing step is to extract salient features from the raw data. Better features produce better predictions. This brief proposes a systematic method to discover such features, utilising context-free grammars to generate a large candidate set from which a subset is selected. Expert knowledge can be incorporated in the grammar, and the selection process avoids redundant and irrelevant features. The readers can use the techniques in this brief to a wide range of machine learning applications to represent complex feature dependencies explicitly when machine learning cannot achieve this by itself.

The utility of grammar-based feature selection is demonstrated using examples involving stock market indices, peak electricity load and foreign exchange client trade volume. The application of machine learning techniques to predict stock market indices and electricity load is not a recent development. Yet it continues to attract considerable attention due to the difficulty of the problem, which is compounded by the nonlinear and non-stationary nature of the time-series. In addition to these two popular applications, we have also demonstrated the application of the technique to foreign exchange client trade volume time-series (irregular time-series). This unique application particularly focuses on hedging the risk exposure of a bank by predicting client trade volumes and directions.

This brief was produced as a result of the research carried out at the Computer Engineering Laboratory (CEL), University of Sydney, Australia with the collaboration of Westpac Banking Corporation. Anthony Mihirana De Silva was funded by the University of Sydney International Scholarships (USydIS) scheme. We also gratefully acknowledge the support of the Australian Research Council's Linkage Projects funding scheme (Project number LP110200413) and Westpac Banking Corporation.

We are indebted to Dr. Barry Flower who has been a long-time companion of our research and a constant source of insights. Dr. Ahmed Pasha and Dr. Richard Davis provided valuable ideas and discussions regarding feature selection, and Farzad Noorian's help with crucial ideas and implementation has greatly improved the quality of this work.

The University of Sydney, December 2014   Anthony Mihirana De Silva
                                                           Philip H.W. Leong

Contents

Chapter 1
Introduction

Abstract A time-series is a collection of data points recorded at successive points in time and are a common occurrence in a diverse range of applications such as finance, energy, signal processing, astronomy, resource management and economics. Time-series prediction attempts to predict future events/behaviour based on historical data. In this endeavour, it is a considerable challenge to capture inherent nonlinear and non-stationary characteristics present in real-world data. Success, or otherwise, is strongly dependent on a suitable choice of input features which need to be extracted in an effective manner. Therefore, feature selection plays an important role in machine learning tasks.

Keywords Time-series prediction · Applied machine learning · Importance of feature selection · Feature generation · Machine learning hypothesis

1.1 Time-Series Prediction

Rapid advances in data collection technology have enabled businesses to store massive amounts of data. Data mining algorithms are used to analyse such stored data to reveal previously unknown strategic business information in the form of hidden patterns and trends which are not at first apparent.

Time-series have many application domains and are commonly utilised in a diverse range of disciplines including engineering, economics, geography, mathematics, physics, marketing and social sciences. Time-series prediction is a form of data mining that predicts future behaviours by analysing historical data. Share prices, foreign exchange rates and yearly company profits are examples of financial time-series, while temperature, sunspot activity and daily average rainfall are examples arising from physical phenomena. Figure 1.1 depicts some examples of time-series: annual number of sunspots, well-studied EUNITE daily electricity load [1] and closing price of the Standard & Poor's 500 (GSPC) stock index.

A time-series is a sequence of vectors (or scalars) recorded at successive points in time. A vector sequence produces a multivariate time-series, and a scalar sequence

© The Author(s) 2015
A.M. De Silva and P.H.W. Leong, *Grammar-Based Feature Generation*
for Time-Series Prediction, SpringerBriefs in Computational Intelligence,
DOI 10.1007/978-981-287-411-5_1

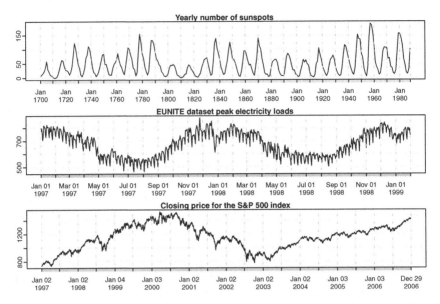

Fig. 1.1 Three examples of real-world time-series. The *top* plot shows the annual number of sunspots, middle is a portion of the EUNITE electricity load benchmark, and *bottom* is the closing price of the Standard & Poor index

produces a univariate time-series. The sequence $x(t_0)$, $x(t_1)$, ..., $x(t_{i-1})$, $x(t_i)$, $x(t_{i+1})$, ... is a univariate time-series of the recorded variable x.

Time-series can be recorded at regular or irregular intervals, and the recorded values can be discrete or continuous. A series of events occurring randomly in time is a special type of time-series known as a point process, e.g. the dates of major earthquakes. In this brief, the time-series of interest are financial time-series of daily stock index prices, daily peak electricity load time-series and foreign exchange client trade volume time-series (irregular).

The objective of a time-series prediction task at time t is to estimate the value of x at some future time, $\hat{x}[t+s] = f(x[t], x[t-1], \ldots, x[t-N])$, $s > 0$ is called the horizon of prediction, e.g. for one-step ahead predictions $s = 1$. Figure 1.2 shows the prediction of a time-series using autoregressive integrated moving average (ARIMA)

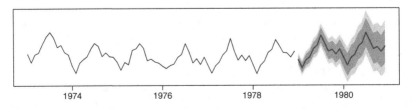

Fig. 1.2 Time-series prediction: predicting the daily values using ARIMA

model. The years 1973–1979 are the historical data, and the years 1980–1981 are the predictions. The horizon of prediction is 2 years of daily values ($s = 730$ days), and the prediction confidence limits are also shown.

Hyndman and Athanasopoulos discuss three general model-based approaches to time-series prediction: explanatory models, time-series models and mixed models in reference [2].

Explanatory models use predictor variables commonly referred to as features in the machine learning (ML) terminology. Suppose that the prediction task is to predict the AUD/USD exchange rate ($E_{AUD/USD}$) for the next month. An explanatory model can be of the form, $E_{AUD/USD} = f$ (Australian interest rate, US interest rate, strength of economy, gross domestic product, foreign relations, error).

If $E_{AUD/USD}$ is denoted by E, a simple *pure time-series model* of the form $E_{t+1} = f(E_t, E_{t-1}, E_{t-2}, E_{t-3}, \ldots, \text{error})$ can be constructed. Such models only use information from the variable to be predicted and makes no attempt to discover the factors affecting its behaviour, i.e. it will extrapolate trend and seasonal patterns but ignore all other external covariates such as Australian interest rate, US interest rate, strength of economy, gross domestic product and foreign relations.

A mixed model can be formed as a combination of explanatory and pure time-series models, e.g. $E_{t+1} = f(E_t, E_{t-1}, E_{t-2}, E_{t-3}, \ldots,$, Australian interest rate, US interest rate, strength of economy, gross domestic product, foreign relations, error).

Pure time-series models such as ARIMA and exponential smoothing state space model (ETS) are in much wider use than explanatory models because it is difficult to understand and accurately model the complex relationships between the explanatory variables and the target variable. Many time-series models are based on linear methods [3] in which the output variable depends linearly on its own previous values. Real-world time-series are often nonlinear and non-stationary. Nonlinear approaches such as nonlinear autoregressive processes, bilinear models and threshold models are widely used for time-series modelling. The generalised autoregressive conditional heteroskedasticity (GARCH) model is another nonlinear time-series approach used to represent the changes of variance over time (heteroskedasticity).

The drawback of model-based approaches is that usually a priori assumption of the underlying distribution of data is required for model parameter estimation. ML techniques can alleviate this issue and cope with the inherent nonlinear and non-stationary nature of real-world time-series.

1.2 Machine Learning Techniques for Time-Series Prediction

Machine learning is concerned with teaching computers to make predictions or behaviours based on information extracted from data. Supervised ML methods have been extensively applied to a range of time-series prediction problems including financial [4–7], energy markets [8–10], control system/signal processing and resource management (see [11, 12] for a comprehensive list of applications). The inherent nonlinear and non-stationary nature of real-world time-series makes ML methods, also referred to as learners, more appealing than model-based approaches [13–16].

Time-series prediction using ML can be treated as a supervised learning task. The learner is presented with the training samples $(\mathbf{x_1}, y_1), (\mathbf{x_2}, y_2), \ldots, (\mathbf{x_n}, y_n)$, where $\mathbf{x}_j \in \mathbb{R}^N$ is a vector of features. For a regression task, $y_j \in \mathbb{R}$ is the target output, and for a classification task, $y_j \in \mathbb{Z}$ (and finite) is the class label. The learner's objective is to find structure in the training samples to infer a general function (a hypothesis) that can predict y_k for previously unseen $\mathbf{x}_k$. The generalisation is achieved by searching through a hypothesis space $\mathcal{H}$, for a hypothesis that best fits the training samples. In a binary classification example, the target hypothesis $h : \mathbb{Z} \rightarrow \{0, 1\}$ is a discriminant boundary that maximises the classification accuracy, for support vector and neural network regression, and the hypothesis $h : \mathbb{R}^n \rightarrow \mathbb{R}$ is a function that minimises the root mean squared error, and so on (see also Sect. 1.3).

Real-world time-series prediction problems are often complex and have interdependencies that are not clearly understood. The distinction of ML algorithms from the time-series model-based approaches is the aspect of learning (training). For a neural network, learning represents the optimisation of the network weights. For a SVM, this is the construction of a hyperplane in a high dimensional space. These optimisations are done using a set of training samples, i.e. known outcomes of the problem for given conditions. The final data modelling goal is not to memorise known patterns but to generalise from prior examples, i.e. be able to estimate the outcomes of the problem under unknown conditions. The ML method has to extract from the training samples, something about the distribution to allow it to produce useful predictions in new, unseen cases. A ML task starts with the construction of a data set. Figure 1.3 illustrates the steps involved with the training phase of a ML algorithm.

1. **Preprocessing:** Data preprocessing includes simple operations such as data cleaning, normalisation, standardization and more advanced operations such as feature transformation and feature selection (FS). Preprocessing can be time consuming but has been shown to produce significantly better results due to better representation of the solution hypothesis. Usually, features are considered individually for simple preprocessing tasks. Any outliers are removed, and non-available data samples are replaced with a mean value, closest value, regressed value or the most common value. The features are often scaled to the range $[-1, 1]$ or $[0, 1]$ to ensure that each feature is independently normalised using min–max or z-score

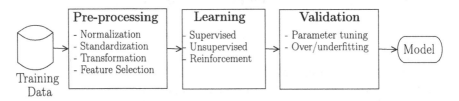

Fig. 1.3 Training phase of a machine learning method

normalisation [17]. The concepts of FS and feature transformation (sometimes referred to as feature construction) are discussed in Chaps. 2 and 4.

2. **Learning:** This is the core aspect of a learning method. Unsupervised learning involves discovering structure in unlabelled data, e.g. clustering algorithms. On the other hand, supervised learning infers a mapping function that maps input features to outputs (or labels), e.g. regression and classification algorithms. A reinforcement learning agent learns by interacting with its environment and observing the feedback of these interactions which in a way mimics human behaviour. It involves trial and error to maximise a certain reward, e.g. in robotics applications.

3. **Validation:** Cross-validation is a way of measuring the performance of a ML method before the algorithm is deployed as a real-world application. It is easy to overfit a model by including too many degrees of freedom. This will lead to small errors in learning but poor generalisation for unseen data. Therefore, validation sets must always be used for model optimisation such as parameter tuning. This process is described in Chap. 5.

1.3 Importance of Feature Selection in Machine Learning

ML algorithms can be viewed as techniques for deriving one or more hypothesis from a set of observations. To facilitate this, the input (observations) and the output (hypothesis) need to be specified in some particular language. As shown in Fig. 1.4, the training samples (observations) ε can be specified using an observation language $\mathscr{L}_\varepsilon$, and a hypothesis $h \in \mathscr{H}$ can be specified by a language $\mathscr{L}_H$. $\mathscr{L}_\varepsilon$ is the notation used to represent the data used for training (training patterns), e.g. in a typical problem, the observation language is a feature-value vector where all observations can be represented using the same set of fixed features. $\mathscr{L}_H$ is the notation used by the learner to represent what it has learned, e.g. in a neural network, the notation is used to represent weights. The performance of supervised ML methods depends strongly on the formalism in which the solution hypothesis is represented using $\mathscr{L}_H$. The features used in $\mathscr{L}_\varepsilon$ and $\mathscr{L}_H$ are identical which means that one way to achieve a better formulation of $\mathscr{L}_H$ is a better representation of $\mathscr{L}_\varepsilon$. Therefore, an important research area is to investigate techniques to expand the feature space used to represent $\mathscr{L}_\varepsilon$ and

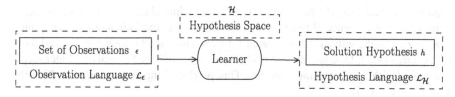

Fig. 1.4 Language terminology (as depicted in [18])

select features that maximise the performance of a particular ML architecture under consideration.

This raises the question "Can we improve the performance of a particular ML architecture by expanding and then reducing the feature space to better formalise the representation of the solution hypothesis?" The main aim of this brief is to introduce and assess the performance of a flexible framework that can transform the initial feature space of $\mathscr{L}_\varepsilon$ to a much larger feature space containing thousands of feature combinations with different parameters and then extract feature subsets from this new space that can produce better predictions, i.e. a better formulation of $\mathscr{L}_H$.

Expert features in time-series prediction applications It was already mentioned that real-world time-series are often nonlinear and non-stationary. For a time-series $x[t]$, the auto-covariance is given by $\gamma(x[t + \tau], x[t]) = cov(x[t + \tau], x[t])$, $\forall\ t$ and lag τ. $x[t]$ is stationary if two conditions are satisfied. The first condition is that $x[t]$ is finite and does not depend on t, i.e. $E(x[t]) = \mu_{x[t]} = \mu < \infty$. The second condition is, for each lag τ, the auto-covariance does not depend on t, i.e. $\gamma(x[t + \tau], x[t]) = \gamma_\tau$. In simple terms, the statistical properties of a stationary time-series are time invariant. In a non-stationary time-series, the distribution of the time-series changes over time causing changes in the dependency between the input and output variables. An effective learning algorithm should have the potential to take this into account. The sliding window (or the rolling window) technique for one-step ahead predictions described in Chap. 5 can be used to create new prediction models when new data become available.

If the time-series to be predicted is denoted by $y[t]$ and the features by $x_1[t]$, $x_2[t]$, ..., $x_k[t]$, each feature is a time-series by itself. In one-step ahead predictions, the objective is to predict the value of $y[t + 1]$ using the previously unseen feature vector $x_1[t], x_2[t], \ldots, x_k[t]$. ML techniques can be applied for time-series prediction as a classification task or a regression task. For regression tasks, the target variable is $y[t]$ or an appropriately smoothed version of $y[t]$ [19]. For classification tasks, a discrete target variable is usually constructed from a continuous target variable $y[t]$. For example, in a daily stock closing price prediction task, if the closing price $C[t]$ has an upward directional change $C[t + 1] - C[t] \geq 0$, $y[t] = +1$, and if the price has a downward directional change $C[t + 1] - C[t] < 0$, $y[t] = -1$, i.e. $y[t] \in \{-1, +1\}$. For both regression and classification, appropriate features need to be utilised.

As an example application, the approach used by kim [20] to predict the daily directional change of the Korean composite stock index (KOSPI) was to use technical indicators as features. The first feature x_1 was the technical indicator moving average convergence divergence (MACD). The feature x_2 was the indicator disparity, and feature x_3 was the indicator RSI. Figure 1.5 shows the target (closing price) and 3 of the features (MACD, disparity and RSI). Reference [20] used a total of 12 technical indicators with 1637 training samples. This produced the feature matrix $\mathscr{X}$ with $k = 12, n = 1{,}637$.

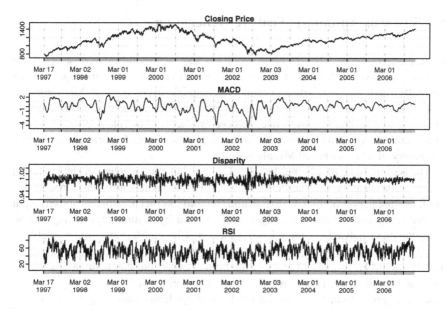

Fig. 1.5 Next day's closing price is the target, and MACD, Disparity and RSI are used as features

$$\mathcal{X} = \begin{pmatrix} x_1[t-n] & x_2[t-n] & \cdots & x_k[t-n] \\ x_1[t-(n+1)] & x_2[t-(n+1)] & \cdots & x_k[t-(n+1)] \\ \vdots & \vdots & \ddots & \vdots \\ x_1[t-2] & x_2[t-2] & \cdots & x_k[t-2] \\ x_1[t-1] & x_2[t-1] & \cdots & x_k[t-1] \end{pmatrix}$$

The ML algorithm was trained with the feature matrix $\mathcal{X}$. Five hundred and eighty-one unseen feature vectors were presented to the model to get the predictions. For example, the unseen vector $x_1[t], x_2[t], \ldots, x_k[t]$ was used to predict the target variable $y[t+1]$, the vector $x_1[t+1], x_2[t+1], \ldots, x_k[t+1]$ was used to predict the target variable $y[t+2]$, and so on. It was necessary to ensure that at any time in predicting the value of $y[t+1]$, the feature matrix had no access to information beyond time t. Such a phenomenon which will lead to unrealistically good predictions is referred to as "peeking" within this brief.

The performance of ML techniques in predicting time-series, among other things, depends on suitable crafting of feature vectors that contain information about the target signal. In most cases, these are selected by experienced users and domain experts. The next section explains the concept of FS using computational methods.

1.4 Towards Automatic Feature Generation

Feature generation enriches the observation language with additional constructed and derived features [21]. A goal of feature generation is to represent feature dependencies explicitly when the ML algorithm cannot accomplish this by itself. A ML algorithm might be able to elucidate hidden dependencies in data by itself, but this can be supplemented in practice by proper feature engineering to enhance performance. This section sheds some light on the relationship between the features and the innate machinery of a kernel-based ML algorithm.

The performance of a kernel-based ML algorithm can be enhanced by (i) kernel optimisations, e.g. type of kernel (only the classical kernels are considered in this brief, i.e. linear, polynomial, Gaussian and sigmoid), kernel parameters and (ii) input feature space (observation language) manipulation by feature engineering. These tasks can be visualised as tuning knobs of a kernel-based ML algorithm available to a human expert (see Fig. 1.6). Ideally, kernel optimisation and feature space manipulation should be concurrent tasks of the human expert but this is infeasible due to the large search space. Therefore, in practice, an appropriate balance is required. For time-series prediction, the additional burden of empirically evaluating parameterised features, e.g. a moving average of a lagged time-series as a feature, and determining for each feature (i) which lags to use, (ii) what look-back periods to use, (iii) what are the moving-widow sizes to use and (iv) what are the best feature combinations to use can have a great impact on ML performance. The framework described in this brief attempts to decouple kernel optimisation and feature engineering. The same philosophy can be extended to non-kernel-based ML algorithms.

The context-free grammar (CFG)-based feature generation framework described in Chap. 4 systematically invokes the defined grammar rules to generate a large feature space. Expert suggested features are expanded by a range of operators (see Fig. 1.7). The grammars are formally defined using the Backus-Naur form (BNF) notation, customizable and organised into grammar families. A user with a better understanding of the time-series under consideration, i.e. an experienced user or a domain expert, can define focused grammars to generate features that are more suitable. Such an approach can produce thousands of features. Therefore, a good feature generation framework should have the inherent capability to scan this large feature space to extract best features for the learner under consideration, i.e. find a better solution hypothesis.

If the solution hypothesis formulated using expert defined features is h_1 and that formulated using a feature subset discovered through the feature generation framework is h_2, which solution hypothesis is better? The thesis of this brief is h_2 is often preferable.

Fig. 1.6 Tuning knobs of a kernel-based ML algorithm

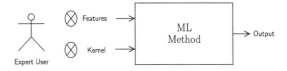

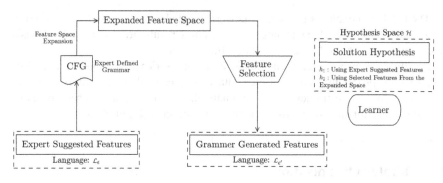

Fig. 1.7 The research question. Which hypothesis is better? h_1 or h_2?

1.5 Contributions to Formalising a Better Hypothesis

This brief describes the implementation of a framework to systematically generate suitable features using CFGs. The notion of grammar families is introduced. Implementation issues affecting the feature space size, feature informativeness, computational time and framework flexibility are studied in detail and ways to overcome them proposed. Some of the grammar families use the concept of *technical indicators* [22], as used in finance. These are formulae that identify patterns and market trends in financial markets which are developed from models for price and volume. Other grammar families to capture information using wavelet transformations, history windows and other operators are also described.

Depending on the configuration of the grammar framework, irrelevant and redundant features can be produced. FS eliminates such features, thereby improving the performance and speed of ML algorithms [23]. Most ML techniques are designed to find the relevant features, but an additional pre-processing step is often required prior to the application of ML techniques to get the best performance. Furthermore, using all of the generated features in a learning task is impractical and leads to overfitting. This brief also compares different FS filters such as information gain, symmetric uncertainty, correlation, maximum relevance minimum redundancy (mRMR), Relief and wrapper techniques such as sequential forward and backward selection and genetic algorithms to investigate the effectiveness of the selected algorithms in mining large feature spaces.

The feature generation framework is extended as a hybrid FS and feature generation framework by using a modified version of GE [24]. GE is described as a convenient way to avoid selective feature pruning. This extended system is flexible in that an expert user can also suggest known feature subsets, e.g. feature subsets that are known to work well, and the system attempts to discover feature subsets that give better predictions. To the best of the author's knowledge, this is the first time GE is applied as a feature discovery technique for time-series prediction. The software was developed using the R programming language and is made publicly available since there is no existing R package for GE (available at [25]). Usage examples of the package are provided in Chap. 3.

The brief also explores financial time-series and electricity peak load prediction using features employed in standard practice. These results are considered as a basis in comparison with the feature generation framework.

Finally, foreign exchange client trade volume time-series prediction is also investigated and the proprietary nature of such data having resulted in little published work. The capability of the framework in automatically identifying appropriate types of features to use for such previously unexplored data sets is highlighted.

1.6 Chapter Organisation

This chapter presented the necessary background in time-series prediction using ML techniques and an overview of the ideas in this brief. Several FS techniques are discussed, and the importance of FS is highlighted in Chap. 2. Grammatical evolution (GE) and associated concepts, context-free grammar (CFG) and genetic algorithms (GAs) are explained in Chap. 3. Chapter 4 contains details on how CFG is used as a feature generation framework and descriptive implementation details of a state of the art feature generation framework. Chapter 5 continues to present how FS is performed on the expanded feature space to mine for "good" features. This chapter also describes how the learner is applied to predict time-series by utilising the sliding window technique, cross-validation and parameter tuning. Chapter 6 presents the results of few case studies and discusses the results, and Chap. 7 draws our conclusions on the results. Wavelet theory is deferred to Appendix A. Some production grammar rule sequences that generate interesting features are provided in Appendix B and supplementary results are in Appendix C.

References

1. EUNITE, World-wide competition within the eunite network (2001), http://neuron-ai.tuke.sk/competition/. Accessed Feb 2001
2. R.J. Hyndman, G. Athanasopoulos, Forecasting: principles and practice, OTexts (2014)
3. C. Chatfield, *The Analysis of Time Series: An Introduction*, vol 59 (Chapman and Hall/CRC, London, 2003)
4. W. Cheng, W. Wagner, C.H. Lin, Forecasting the 30 year US treasury bond with a system of neural networks. J. Comput. Intel. Finan. **4**, 10–16 (1996)
5. J.W. Hall, *Adaptive Selection of US Stocks With Neural Nets*. Trading on The Edge: Neural, Genetic, and Fuzzy Systems for Chaotic Financial Markets (New York, Wiley, 1994), pp. 45–65
6. Z. Huang, H. Chen, C.J. Hsu, W.H. Chen, S. Wu, Credit rating analysis with support vector machines and neural networks:a market comparative study. Decis. Support Syst. **37**(4), 543–558 (2004)
7. F.E.H. Tay, L. Cao, Application of support vector machines in financial time series forecasting. Omega **29**(4), 309–317 (2001)
8. M. Espinoza, J. Suykens, B. De Moor, Load forecasting using fixed-size least squares support vector machines. Comput. Intell. Bioinspired Syst. pp. 488–527 (2005)

9. T. Liang, A. Noore, A novel approach for short-term load forecasting using support vector machines. Int. J. Neural Syst. **14**(05), 329–335 (2004)
10. M. Mohandes, Support vector machines for short-term electrical load forecasting. Int. J. Energy Res. **26**(4), 335–345 (2002)
11. B. Krollner, B. Vanstone, G. Finnie, *Financial time series forecasting with machine learning techniques: a survey.* in European Symposium on ANN: Computational and Machine Learning (2010)
12. N. Sapankevych, R. Sankar, Time-series prediction using support vector machines: a survey. IEEE Comput. Intel. Mag. **4**(2), 24–38 (2009)
13. P.J. Brockwell, R.A. Davis, *Introduction to Time-Series and Forecasting* (Springer, New York, 2002)
14. J. Franke, W. Härdle, C.M. Hafner, *Statistics of Financial Markets: An Introduction* (Springer, New York, 2008)
15. T.L. Lai, H. Xing, *Statistical Models and Methods for Financial Markets*, (Springer, New York, 2008)
16. R.S. Tsay, *Analysis of Financial Time Series*, vol. 543 (Wiley, New York, 2005)
17. S.B. Kotsiantis, D. Kanellopoulos, P.E. Pintelas, Data preprocessing for supervised leaning. Int. J. Comput. Sci. **1**(2), 111–117 (2006)
18. C. Sammut, G.I. Webb, *Encyclopedia of Machine Learning* (Springer, New York, 2011)
19. L.-J. Cao, E.H.F. Tay, Support vector machine with adaptive parameters in financial time series forecasting. IEEE Trans. Neural Netw. 14(6), 1506–1518 (2003)
20. K. Kim, Financial time series forecasting using support vector machines. Neurocomputing **55**(1), 307–319 (2003)
21. O. Ritthof, R. Klinkenberg, S. Fischer, I. Mierswa, (2002) A hybrid approach to feature selection and generation using an evolutionary algorithm. In 2002 U.K. workshop on computational intelligence, pp. 147–154
22. C.D. Kirkpatrick, J.R. Dahlquist, *Technical Analysis: The Complete Resource for Financial Market Technicians* (Safari Books Online, FT Press Financial Times, Essex, 2007)
23. H. Liu, L. Yu, Toward integrating feature selection algorithms for classification and clustering. IEEE Trans. Knowl. Data Eng. **17**(4), 491–502 (2005)
24. M. O'Neill, C. Ryan, Grammatical evolution. IEEE Trans. Evol. Comput. **5**(4), 349–358 (2001)
25. A.M. de Silva, F. Noorian, R package for grammatical evolution (2013), http://cran.r-project. org/web/packages/gramEvol/index.html. Accessed Nov 2013

Chapter 2
Feature Selection

Abstract Feature selection (FS) is important in machine learning tasks because it can significantly improve the performance by eliminating redundant and irrelevant features while at the same time speeding up the learning task. Given N features, the FS problem is to find the optimal subset among 2^N possible choices. This problem quickly becomes intractable as N increases. In the literature, suboptimal approaches based on sequential and random searches using evolutionary methods have been proposed and shown to work reasonably well in practice. This chapter describes the mainstream feature selection technique theories.

Keywords Redundant and irrelevant features · Feature interaction · Feature quality evaluation · Feature selection models · Over-fitting learning models

2.1 Introduction

FS is the process of choosing a subset of features that improves the ML method performance. Formally, for N data samples with M features in each data sample, FS problem is to find from the M-dimensional observation space $\mathbb{R}^M$, a subspace of m features $\mathbb{R}^m$ that best predict the target. This is achieved by reducing the number of features to remove irrelevant, redundant, or noisy information from the feature set.

FS primarily aids in alleviating the curse of dimensionality and speeding up the learning task. In practice, it also helps in optimizing data collection methodologies by identifying which data to collect. Therefore, FS is a key pre-processing step (see Fig. 1.3) to improved predictions.

In a typical ML problem, there is usually an optimal number of features that provide the minimum error (highest accuracy). Because the total number of subspaces is 2^M, finding an optimal feature subset is usually intractable [1] and many problems related to FS have been shown to be NP-hard [2]. Alternatively, many sequential and random searches have been proposed.

© The Author(s) 2015   13
A.M. De Silva and P.H.W. Leong, *Grammar-Based Feature Generation*
for Time-Series Prediction, SpringerBriefs in Computational Intelligence,
DOI 10.1007/978-981-287-411-5_2

2.1.1 Redundant and Irrelevant Features

An irrelevant feature carries no useful information in describing the relationships of the underlying data. However, a feature that is irrelevant by itself may become useful when considered in combination with some other features, e.g. learning the 2-input XOR function given one input is not possible hence one input is irrelevant by itself but the 2 inputs in combination can produce the decision function. The elimination of redundant features is another aspect in FS. This can only be addressed by considering feature subsets and not by individual feature relevance assessment. If highly correlated features are present, individual features may exhibit similar performance to the collective feature subset hence using only the non-redundant features will improve performance. Perfectly correlated variables are truly redundant in the sense that no additional information is gained by adding them, and high variable correlation (or anti-correlation) does not imply the absence of variable complementarity [3].

Redundant features can be handled by dimensionality reduction techniques such as principal component analysis (PCA), independent component analysis (ICA) and kernel principal component analysis (KPCA). Linear techniques such as PCA perform a linear mapping of the data to a lower dimensional space in such a way, the variance of the data in the low-dimensional representation is maximized. KPCA and manifold techniques are able to perform non-linear dimensionality reduction. A good review can be found in Ref. [4].

2.1.2 Feature Interaction

In the presence of multiple interacting features, the correlation of individual features with the target variable may not be significant. However, if such interacting features are considered in combination, they could show a strong relationship to the target variable. Jakulin and Bratko [5] define an interacting subset as an irreducible whole, i.e. removing a variable impedes prediction ability. While Jakulin and Bratko introduce feature selection algorithms to deal with 2-way (one feature and the class) and 3-way (two features and the class) interactions, other algorithms such as INTERACT can detect n-way interactions in a subset. It is questionable if real interacting feature subsets can be efficiently discovered but in our experience, the wrapper techniques discussed later lead to good results at the expense of computational time.

2.1.3 Over-Fitting and the Optimal Number of Features

Given a finite number of training samples with redundant features, the error rate of a typical learner will drop and then increase as the number of features is increased. Hence, in a typical problem, there exists an optimal number of features that provide the minimum error (highest accuracy). It is easy to over-fit a model by including too many degrees of freedom which will lead to poor generalization. The importance of having a separate training set for FS has been investigated for regression and classification [6]. The authors of [6] point out if the same dataset is used for feature selection and the learning task, the relationship between the features and the target variables may appear much stronger than actuality. This is called "feature selection bias". Cross validation is therefore a key aspect in effective FS and validation sets are important for model optimization such as model parameter-tuning and kernel selection (in kernel methods). K-fold cross validation, two-fold cross validation, leave-one-out cross validation and repeated random sampling cross validation are some popular cross validation techniques [7]. Time-series cross validation is slightly different because the data are not independent and, due to the correlations with other observations, leaving out an observation does not remove all the associated information. Cross validation can be used to choose the number of features that provides the best accuracy in the training (in-sample) data (see Chap. 5). Unfortunately, this does not ensure that the same features are good for testing (out-sample) data.

2.1.4 Feature Quality Evaluation

The performance of a ML task before and after feature selection can be evaluated to verify that the FS algorithm has served its purpose. A FS algorithm should not only strive to improve the prediction performance, but also consider other important attributes such as the complexity of the resulting model, interpretability of the model, subset cardinality, computational time for feature selection, scalability and generalization of the resulting model.

2.2 Feature Selection Models

Three feature selection models are considered: filter, wrapper and embedded. Filter models are independent of the learning algorithm and hence solely data dependent. Wrappers optimize with the specific learning function in the loop, assessing the performance of different feature subsets based on a cost function. Embedded models are built into the learning algorithm itself, for example, via a specific way of determining the weights in a neural network or a modified penalty term in a kernel based model function. Selected techniques from each category are reviewed in this brief. Feature ranking assigns a weight for each feature and feature subset selection evaluates different feature combinations. In general, filter models are used as feature ranking strategies and wrapper and embedded models are used as feature subset selection strategies.

2.2.1 Filter Models

A filter model makes use of intrinsic characteristics in the data for feature ranking. The feature ranking criterion can be based on dependency measures, distance measures and consistency measures. In this research, the commonly used filters include the information gain (dependency), maximum-relevance-minimal-redundancy (dependency), Pearson's correlation (dependency) and the Relief algorithm (distance). By evaluating these measures between the feature and the target variable the features can be assigned a weight and they can be ranked.

1. **Correlation**: Also called similarity measures or dependency measures. In time-series prediction tasks on stationary data, tests such as a Granger causality test can be used as a tool to aid in deciding whether one time-series can be useful in predicting another. If X and Y are two time-series, X is said to Granger-cause Y if Y can be better predicted using the histories of both X and Y than it can by using the history of Y alone. This is, in fact, a form of feature selection.
   A good individual feature is highly correlated with the target so correlation measures can be used for ranking or subset selection. As an example, for a candidate feature $x_i \in X$ and regression target Y, the Pearson correlation coefficient is given by $\rho(x_i, Y) = \dfrac{cov(x_i, Y)}{\sqrt{\sigma(x_i)\sigma(Y)}}$ where cov is the covariance, and σ the variance.
   Direct feature-target correlation by itself is however considered as a poor technique because a high feature-target correlation can still be an irrelevant feature. These are known as spurious correlations.

2. **Information Gain**: Information theoretic criteria can be used in place of correlation. Information (the negative of entropy) contained in a discrete distribution of feature X is given by,

$$H(X) = -\sum_i p(x_i) \log_2 p(x_i) \tag{2.1}$$

where the x_is are the discrete feature values and $p(x_i)$ is its probability. Continuous features are either discretized, or integration instead of summation is performed by fitting a kernel function to approximate the density of the feature X. Information embedded in the joint distribution is provided by,

$$H(Y, X) = -\sum_i \sum_j p(y_j, x_i) \log_2 p(y_j, x_i) \tag{2.2}$$

where $p(y_j, x_i)$ is the joint probability. Mutual information (MI) provides a good measure of feature importance. MI which is calculated as $MI(Y, X) = H(Y) + H(X) - H(Y, X)$ is equal to the Kullback-Leibler divergence given by,

$$MI(Y, X) = -\sum_{i,j} p(y_j, x_i) \log_2 \frac{p(y_j, x_i)}{p(y_j)p(x_i)} \tag{2.3}$$

A feature is more important if the mutual information MI(Y, X) between the target and the feature distributions is larger. Information gain is a similar criterion where $IG(Y, X) = H(Y) H(Y|X)$ is the difference between information contained in the class distribution $H(Y)$, and information after the distribution of feature values is taken into account, i.e. the conditional information $H(Y|X)$.

3. **Maximum-relevance-minimal-redundancy (mRMR)**: mRMR is a feature selection technique that attempts to constrain features to a subset which are mutually as dissimilar to each other as possible, but as similar to the classification variable as possible. Maximum relevance features are assessed by the following equation where S is a feature set of m features ($|S| = m$); x_i is an individual feature; c is the target class and $IG(x_i, c)$ is the difference between information contained in the target distribution $H(c)$, and conditional information $H(c|x_i)$ as discussed above:

$$\max_{S,c} D = \frac{1}{|S|} \sum_{x_i \in S} I(x_i, c). \tag{2.4}$$

Although such features are highly relevant, they could also be highly redundant. A criterion to select mutually exclusive features can be written as:

$$\min_{S} R = \frac{1}{|S|^2} \sum_{x_i,x_j \in S} I(x_i, x_j) \tag{2.5}$$

The "maximum-relevance-minimal-redundancy" criterion combines the above two goals by optimizing the cost function:

$$\max_{D,R} \Phi = D - R \tag{2.6}$$

Starting from an initial feature set S_{m-1} with $m - 1$ features, the objective is to select the mth feature from the remaining set $\{X - S_{m-1}\}$ by maximizing $\Phi(\cdot)$. This is done incrementally via Eq. 2.7 to find the solution.

$$\max_{x_j \in X - S_{m-1}} \left[I(x_j, c) - \frac{1}{m - 1} \sum_{x_i \in S_{m-1}} I(x_j, x_i) \right] \tag{2.7}$$

4. **Relief Score**: Distance based filter models such as Relief ranks features that distinguish classes based on how well a feature can separate classes. The original Relief algorithm, proposed by Kira and Rendell [8], is a two-class filtering algorithm for features normalized to [0, 1]. Each feature is initially assigned a zero weight. An A-dimensional training example R is chosen randomly and the Euclidean distance to all other instances calculated. Denote the nearest hit in the same same class H, and the nearest miss in a different-class M. Since a good feature

R[A] should be able to separate class values, it should have a small distance to H and a large distance to M. Hence W[A] is adjusted to reward good features and penalize poor ones. The final selection of features is made by selecting those large W[A], i.e. those that exceed a given threshold. The pseudo code for the Relief algorithm is given below:

Algorithm 1 Relief Algorithm
___
1: set all weights W to 0;
2: **for** i := 1:m do **do**   ▷ m is the training set size
3:      randomly select an instance R;
4:      find nearest hit H and nearest miss M;
5:      **for** A := 1:a do **do**   ▷ a is the dimensionality of the feature vector
6:          W[A] := W[A] - diff(R[A],H[A]) + diff(R[A],M[A]);
7:      **end for**
8: **end for**

Different diff functions can be used for discrete e.g. diff$(x, y) = 0$ if x and y are in the same class, 1 otherwise, and continuous feature values, e.g. diff$(x, y) = (x - y)^2$. RReliefF (regression ReliefF) was developed later for regression problems. Since regression applies to continuous variables, RReliefF uses a probability measure which is modeled by target variable values of instances belonging to different classes. Two disadvantages associated with Relief algorithms are (i) they are computationally expensive and (ii) they may fail to remove redundant features.

5. **Consistency-based filters**: Consistency measures attempt to find a minimum number of features that distinguish between the classes as consistently as the full set of features. An inconsistency arises when multiple training samples have the same feature values, but different class labels. Dash [9] presents an inconsistency-based FS technique called Set Cover. An inconsistency count is calculated by the difference between the number of all matching patterns (except the class label) and the largest number of patterns of different class labels of a chosen subset. If there are n matching patterns in the training sample space and there are c1 patterns belonging to class 1, c2 patterns belonging to class 2 and c3 patterns belonging to class 3, and if the largest number is c2, the inconsistency count would be defined as n − c2. By summing all the inconsistency counts and averaging over the size of the training sample size, a measure called the inconsistency rate for a given subset is defined. This is advantageous because (i) a feature subset can be evaluated in linear time (ii) can eliminate both irrelevant and redundant features (iii) some noise is reflected as a percentage of inconsistencies and (iv) unlike measures like correlation, information theoretic criterion and distance, this is a multivariate measure. Despite these advantages, it can be shown that Set Cover has a tendency to choose features that are highly correlated with the target, although in reality they can be irrelevant. For large dimensional feature spaces, filtering can offer reduced computational time compared to wrappers but this is dependent

on the sample size and the number of features. Other methods can also be applied following filtering. However, the question remains on which relevance measures should be used in a particular application.

In the presence of multiple interacting features, the individual feature relationship with the target can be insignificant but if such interacting features are considered in combination, a strong relationship to the target variable *may* be identified. In other words, "the *m* best features are not the best *m* features" [10]. Many efficient FS algorithms assume feature independence, ignoring the interacting features. Reference [5] defines an interacting feature subset as an irreducible whole: "a whole is reducible if we can predict it without observing all the involved variables at the same time". Therefore, different feature combinations are often evaluated using wrapper models. For large dimensional feature spaces, a filter model is usually applied first to reduce the feature space dimensionality followed by the application of a wrapper model.

2.2.2 Wrapper Models

A wrapper model involves a feature evaluation criterion, a search strategy and a stopping criterion as depicted in Fig. 2.1. Unlike the filter model, a wrapper model utilizes the performance of the learning algorithm as the evaluation criterion. This ensures that the selected feature subset is well matched to the ML algorithm. Wrappers are computationally expensive since they require the ML algorithm to be executed in every iteration, and are usually used for low dimensional datasets. Search strategies generate different feature combinations to traverse through the feature space (generating feature subsets for evaluation). A widely used stopping criterion is to stop generating and evaluating new feature subsets when adding or removing features does not make any performance improvements [2]. The sequential steps of a wrapper model with X being the training data, A the learning algorithm, and S_0 the initial subset of features as given in Ref. [2] is provided in Algorithm 2.

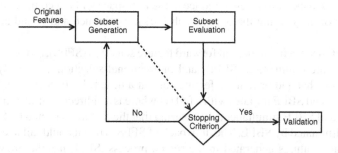

Fig. 2.1 A general wrapper model [2]

Algorithm 2 Wrapper Approach

Input
  X ▷ Training data with N features, $|X| = n$
  N   ▷ Number of features (feature space dimensionality)
  n   ▷ Number of training examples
  A   ▷ Learning Algorithm
  SC   ▷ Stopping criterion
  S_0 ▷ Initial subset of features

Output S_{best} ▷ Selected feature subset

  $S_{best} \leftarrow S_o$ ▷ Assign the current best feature subset as the initial subset
  $\gamma_{best} \leftarrow eval(S_0, X, A)$
  repeat
    $S \leftarrow SearchStrategy(X)$ ▷ Generate next subset to be evaluated
    $\gamma \leftarrow eval(S, X, A)$
    if $\gamma \geq \gamma_{best}$ **then**
      $\gamma_{best} \leftarrow \gamma$
      $S_{best} \leftarrow S$
    end if
  until SC is reached **return** S_{best}

Search strategies generate different feature combinations to traverse through the feature space (generating feature subsets for evaluation).

For N features there exist 2^N potential subsets. For even modest values of N, an exhaustive search over this huge space is intractable. The feature search therefore plays an important role in wrappers. Exponential, sequential and randomized searches are described next.

2.2.2.1 Feature Search

1. **Exponential search**: An exponential search returns the optimal subset. Although the order of the search space is $\mathcal{O}(2^N)$, the search need not be exhaustive, i.e. heuristics can be introduced to reduce the search space without compromising the chances of discovering the optimal subset [2], e.g. branch and bound and beam search.
2. **Sequential search**: Sequential forward feature selection (SFFS), sequential backward feature elimination (SBFE) and bidirectional selection are greedy search algorithms that add or remove features one at a time [2]. SFFS initiates with an empty set and SBFE initiates with a full set whereas a bidirectional search initiates a SFFS and SBFE simultaneously ensuring that the features selected by SFFS are never eliminated by SBFE. The drawback of SFFS is that it could fail to eliminate redundant features generated in the search process. SBFE has the drawback of not being able to re-evaluate feature usefulness together with other features once

a feature is removed. Plus-L minus-R selection (LRS) search attempts to resolve these issues. This can be understood by the sequential forward search algorithm, which is a simple greedy search, as described by the following pseudo code.

Algorithm 3 SFFS algorithm
___
1: Start with the empty set $Y_0 = \{\varnothing\}$
2: Select the next best feature $x^+ = \arg\max x \in Y_k J(Y_k + x^+)$
3: Update $Y_{k+1} = Y_k + x^+$; $k = k + 1$
4: Go to 2

In the worst case every feature might be selected in which case there will be N features to select from in the first step, $N - 1$ in the second step and so on. Therefore there will be a total of $\sum_{i=1}^{n} i$ feature assessments in the worst case, and is hence $\mathcal{O}(N^2)$.

3. **Randomized search**: In practical applications, the feature space may contain thousands of features, e.g. bioinformatics, text, natural language processing applications and the case studies in this brief. In such applications, it is not feasible to search the entire feature space. Randomized search trades off optimality of the solution for efficiency by searching only a sampled portion of the feature space. GAs have been used as a guided randomized FS technique [11, 12] and is discussed in depth in Sect. 3.2.

Table 2.1 compares the exhaustive, sequential and randomized feature search techniques in terms of time complexity, advantages and disadvantages.

2.2.3 Embedded Models

In contrast to filter and wrapper models, embedded models do not separate learning from FS. For example, support vector machines (SVMs) that inherently incorporate a learning model of loss plus penalty can be equipped with embedded feature selection when a L_1 norm on the weight vector is considered. This is explained below.

Table 2.1 Search criterion comparison

Method	Complexity	Advantages	Disadvantages
Exhaustive	$\mathcal{O}(2^N)$	High accuracy	Computationally expensive
Sequential	$\mathcal{O}(N^2)$	Simple	Less flexible with backtracking
Randomized	$\mathcal{O}(N \log N)$	Users can select between accuracy and speed, avoids being trapped in local optima	Low accuracy

For a given training data set of (x_i, y_i) where the input $x_i \in \mathbb{R}$ and the output $y_i \in \{-1, +1\}$, the SVM finds a hyper-plane that separates the two classes by maximizing the distance to the closest point from either class. For a problem where the two classes cannot be cleanly separated, this search for the optimal hyperplane can be viewed as the optimization problem: $\max_{\beta_0, \beta_1} \frac{1}{\|\beta\|_2^2}$ subject to, $y_i(\beta_0 + x_i^T \beta) \geq 1 - \xi_i, \xi_i \geq 0,$ $\sum_{i=1}^n \xi_i \leq B$ where ξ's are the slack variables associated with the overlap between the two classes, B is the tuning parameter that is responsible for controlling this overlap and β_0, β are the decision boundary variables. Although the problem shown is for binary classification, SVM has also been extended to multi-class classification and regression problems.

The optimization problem above is equivalent to the standard SVM model with a L_2 norm in the form of loss + penalty,

$$\min_{\beta_0, \beta_j} \left[\sum_{i=1}^n 1 - y_i \left(\beta_0 + \sum_{j=1}^q \beta_j h_j(x_i) \right) \right] + \lambda \|\beta\|_2^2 \qquad (2.8)$$

where λ is the parameter which balances the tradeoff between the loss and penalty, n is the number of training samples and q is the dimensionality of the solution space $\mathbb{R}^q$. The classification rule is given by $(\beta_0 + x^T \beta)$. The $h_j(x)$s are usually chosen to be the basis functions of a reproducing kernel Hilbert space (RKHS) which allows the use of the kernel trick to transform the original feature space to a higher dimension [13]. There are specific conditions that should be satisfied in choosing a kernel such as the inner product being positive semi-definite, but we omit the details here. The loss term in the above equations is called the hinge loss, and the penalty is called the ridge penalty which is also used in ridge regression. The ridge penalty shrinks the fitted coefficients β's to approach zero.

Zhu et al. [13] proposed to use the L_1 norm, instead of the standard L_2 norm which will give the equivalent Lagrange version of the optimization problem,

$$\min_{\beta_0, \beta_j} \left[\sum_{i=1}^n 1 - y_i \left(\beta_0 + \sum_{j=1}^q \beta_j h_j(x_i) \right) \right] + \lambda \|\beta\|_1 \qquad (2.9)$$

This is known as the lasso penalty which also shrinks β's towards zero. However, by making λ sufficiently large, some of the coefficients β can be made exactly zero. Therefore, the lasso penalty introduces a better form of integrated feature selection that is not observed when using the ridge penalty.

Ng [14] draws the following conclusions in regularization using L_1 and L_2 norms. (i) L_1 outperforms L_2 regularization for logistic regression when there are more irrelevant dimensions than training examples, e.g. n = 100, and a large number of basis functions, e.g. q = 1,000 (translates to features). This is attributed to the fact that in a sparse scenario, only a small number of true coefficients (β_js) are non-zero hence the lasso penalty works better than the ridge penalty. (ii) L_2 regularization classifies poorly for even a few irrelevant features (iii) Poor performance of L_2 regularization is linked to rotational invariance and (iv) Rotational invariance is shared by a large class of other learning algorithms. These other algorithms presumably have similarly bad performance with many irrelevant dimensions, especially when there are many features than samples. SVM-RFE (Recursive Feature Elimination) is an embedded FS method proposed in [14] for gene selection in cancer classification. If a user wishes to select a subset of only m features, nested subsets of features are selected in a sequential backward elimination manner, which starts with all the features and removes one or more feature at a time (greedy backward selection). At each iteration, the feature that decreases the margin the least will be eliminated until only m features remain. Many other embedded feature selection techniques for different learning algorithms exist but we have specifically focused on SVM embedded models in this brief.

Ensemble Learning Ensemble methods combine multiple learning algorithms, and numerous studies show that they can improve the performance over a single learner. There are two variations of ensemble learning; parallel and serial [15]. A parallel ensemble method combines independently constructed learners from the same dataset, e.g. each learner is trained on a different bootstrap sample (a sample drawn with replacement) from the training dataset. When the learners show diverse performance, i.e. different learners show different errors on unseen data, errors cancel out to produce better results than a single learner. For example, random forests (RFs) are a parallel ensemble technique which reduces error variance. In serial ensembles, the new learner is based on the previously built leaner, i.e. the final model is built in a forward stage by stage pattern, e.g. Adaboost, Gradient tree boosting. For example, in Adaboost, a sequence of weak learners i.e., models that are only slightly better than random guessing, such as small decision trees or decision stumps are trained on repeatedly reweighted data samples.

References

1. R.Kohavi, G.H. John, Wrappers for feature subset selection. Artif. intell. **97**(1), 273–324 (1997)
2. H. Liu, Y. Lei, Toward integrating feature selection algorithms for classification and clustering. IEEE Trans. Knowl. Data Eng. **17**(4), 491–502 (2005)
3. I. Guyon, An introduction to variable and feature selection. J. Mach. Learn. Res. **3**, 1157–1182 (2003)
4. L.J. Cao, K.S. Chua, W.K. Chong, H.P. Lee, Q.M. Gu, A comparison of pca, kpca and ica for dimensionality reduction in support vector machine. Neurocomputing **55**(1), 321–336 (2003)
5. A. Jakulin, I. Bratko, Testing the significance of attribute interactions. in *Proceedings of the Twenty-first International Conference on Machine Learning, ICML '04* (ACM, New York, 2004), p. 52
6. S. K. Singhi, H. Liu, Feature subset selection bias for classification learning. in *Proceedings of the 23rd International Conference on Machine Learning* (ACM, New York, 2006), pp. 849–856
7. I. Guyon, S. Gunn, M. Nikravesh, L. Zadeh, Feature extraction, *Foundation and applications* (Springer, Science and Business Media, 2006)

8. K. Kira, L.A. Rendell, A practical approach to feature selection. in *Proceedings of the 9th International Workshop on Machine Learning, ML92* (Morgan Kaufmann Publishers Inc., San Francisco, 1992), pp. 249–256

9. M. Dash, Feature selection via set cover. in *Knowledge and Data Engineering Exchange Workshop, 1997. Proceedings*, pp. 165–171 (1997)

10. H. Peng, F. Long, C. Ding, Feature selection based on mutual information criteria of max-dependency, max-relevance, and min-redundancy. Pattern Anal. Mach. Intell., IEEE Trans. 27(8):1226–1238 (2005)

11. K.J. Cherkauer, J.W. Shavlik, Growing simpler decision trees to facilitate knowledge discovery, in KDD'96, pp. 315–318 (1996)

12. H. Vafaie, K. De Jong, Genetic algorithms as a tool for restructuring feature space representations. in *Proceedings of 7th International Conference on Tools with Artificial Intelligence*, pp. 8–11 (1995)

13. J. Zhu, S. Rosset, T. Hastie, R. Tibshirani, 1-norm support vector machines. Adv. Neural Inf. Process. Syst. **16**(1), 49–56 (2004)

14. A.Y. Ng, Feature selection, l1 versus l2 regularization, and rotational invariance, in *Proceedings of the Twenty-first International Conference on Machine Learning* (ACM, New York, 2004), p. 78

15. I. Guyon, A. Elisseeff, An introduction to variable and feature selection. J. Mach. Learn. Res. **3**, 1157–1182 (2003)

Chapter 3
Grammatical Evolution

Abstract Grammatical Evolution (GE) involves a mapping from a bit string to a program/expression via a user-defined grammar. By performing an unconstrained optimisation over a population of the strings, programs and expressions which achieve a desired goal can be discovered. A hybrid FS and feature generation algorithm using a modified version of GE is introduced. This algorithm serves as the inherent feature selection mechanism in the feature generation framework. The chapter also describes a R package which implements GE for automatic string expression generation. The package facilitates the coding and execution of GE programs and supports parallel execution.

Keywords Genetic algorithms · Context-free grammar · Evolutionary strategies for feature selection · Grammatical evolution · R gramEvol package

3.1 Introduction

Grammatical evolution (GE) can be used to evolve complete programs in an arbitrary language and like other evolutionary programming techniques such as genetic programming (GP) and evolves a population towards a certain goal. The difference between GE and GP is that GE applies evolutionary operators on binary strings which are then converted using the defined grammar to the final program or expression; on the other hand, GP directly operates on the actual program's tree structure [1]. In GE, a suitable grammar developed for the problem at hand is first specified in Backus–Naur form (BNF). This chapter proposes a hybrid FS and feature generation algorithm using a modified version of GE. In order to describe GE, genetic algorithms (GAs) and context-free grammar (CFG) are brought in the next sections.

3.2 Genetic Algorithms

A number of evolutionary approaches have been applied to FS. An initial population is created and evolved to traverse through the feature space. The best chromosomes of the population are selected using different criteria, and the ML algorithm performance

© The Author(s) 2015
A.M. De Silva and P.H.W. Leong, *Grammar-Based Feature Generation
for Time-Series Prediction*, SpringerBriefs in Computational Intelligence,
DOI 10.1007/978-981-287-411-5_3

using the feature subset represented by a chromosome is used to assess its fitness. GAs have been used as a wrapper technique, thus introducing a search mechanism to avoid enumerating the entire space. A simple approach is to encode features in the chromosome, e.g. the chromosome 01001000 could mean that the 2nd and 5th features are selected [2, 3]. Huang and Wang [4] used a chromosome with parameters of a SVM (the penalty parameter C and the kernel parameter γ). The resultant solution provided the best features and the appropriate SVM parameters. Genetic programming (GP) for FS has been used where a GP-based online FS algorithm which evolved a population of classifiers was used to choose accurate classifiers with the minimum number of features [5]. Multi-population GP has also been used in [6] to develop a method for simultaneous feature extraction and selection.

Canonical Genetic Algorithms Canonical GA operates on n-tuples of binary strings b_i of length l. The n-tuple is termed the population, and the bits of each string are considered to be genes of a chromosome. A chromosome is represented as a binary string, with each consecutive group of k-bits creating a *codon*. A group of codons is called a *gene*. A chromosome may contain more than one gene. The operations performed on the population are selection, crossover and mutation as illustrated in Fig. 3.1.

The initial population is created by randomly setting each bit of a chromosome. The chromosomes are then evaluated based on a given fitness function $\phi(\cdot)$, where each chromosome represents a solution to the formalised problem. The objective function $\phi(b_i)$ gives the fitness score of the chromosome which has to be maximised. The better scoring chromosomes are deemed to have the best genes; hence, they are retained and the low scoring chromosomes are discarded from the population and replaced with new chromosomes. Top scoring chromosomes are termed elite chromosomes. The probability for a chromosome b_i to be selected for recombination is proportional to the relative fitness score given by $\phi(b_i)/\sum_{j=0}^{n}\phi(b_j)$.

Fig. 3.1 The evolutionary process and associated operators

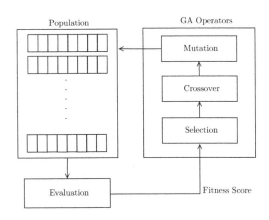

The crossover is performed on 2 randomly selected chromosomes. In canonical GA, the traditional crossover operator is the single-point crossover. A crossover point in the binary string is chosen at random, and the string sections of the two parents are exchanged. The mutation operator randomly flips single bits on a specific chromosome under a defined mutation probability. Mutation is necessary to maintain genetic diversity from one generation of a population to the next. The evolutionary process is repeated until a given termination criterion is satisfied. Integer GAs which are an extension of canonical GAs are explained in Sect. 5.3.2.

Since its introduction by Holland in 1975, other techniques and evolutionary algorithms have been proposed to extend canonical GA. For example, to facilitate complex data representation, GA is often implemented with integer or floating point codons and evolutionary operators are applied directly to the codons instead of the underlying bit string. This method also takes the advantage of the architecture of modern processors to speed-up computation. For a review of other GA techniques, readers are referred to a survey [7].

3.3 Context-Free Grammar

A CFG is a simple mechanism to generate patterns and strings using hierarchically organised production rules [8]. Using the Backus–Naur form (BNF) formal notation [9], a CFG can be described by the tuple $(\mathscr{T}, \mathscr{N}, \mathscr{R}, \mathscr{S})$ where $\mathscr{T}$ is a set of terminal symbols and $\mathscr{N}$ is a set of non-terminal symbols with $\mathscr{N} \cap \mathscr{T} = \varnothing$. The non-terminal symbols in $\mathscr{N}$ and terminal symbols in $\mathscr{T}$ are the lexical elements used in specifying the production rules of a CFG. A non-terminal symbol is one that can be replaced by other non-terminal and/or terminal symbols. Terminal symbols are literals that symbols in $\mathscr{N}$ can take. A terminal symbol cannot be altered by the grammar rules $\mathscr{R}$, and a set of relations (also referred to as production rules) in the form of $\mathscr{R} \rightarrow \alpha$ with $\mathscr{R} \in \mathscr{N}, \alpha \in (\mathscr{N} \cup \mathscr{T})$. $\mathscr{S}$ is the start symbol $\mathscr{S} \in \mathscr{N}$. If the grammar rules are defined as $\mathscr{R} = \{x \rightarrow xa, x \rightarrow ax\}$, a is a terminal symbol since no rule exists to change it, whereas x is a non-terminal symbol. A language is context-free if all of its elements are generated based on a context-free grammar. If $\mathscr{S}$ is the starting symbol and we define a CFG, $\mathscr{T} = \{a, b\}$, $\mathscr{N} = \{\mathscr{S}\}$ and $\mathscr{R} = \{\mathscr{S} \rightarrow a\mathscr{S}b, \mathscr{S} \rightarrow ab\}$, $\mathscr{L} = \{a^n b^n | n \in \mathbb{Z}^+\}$ is a context-free language. An example grammar in BNF notation is given in Table 3.1.

3.4 Generating Features Using GE

In this section, the grammar in Table 3.1 is used to demonstrate how features can be generated using GE. Consider the chromosome [166|225|180|132|187|219|179|249] in which the integer numbers represent codon values of 8 bits, i.e. the chromosome is 64 bits in length. The codon values are used to select production rules from the example grammar definition to generate features. The usual mapping function used

Table 3.1 An example grammar in BNF notation

$\mathcal{N} = \{expr, op, coef, var\}$

$\mathcal{T} = \{\div, \times, +, -, V1, V2, C1, C2, (,)\}$

$\mathcal{S} = <expr>$

Production rules: $\mathcal{R}$

$\langle expr \rangle$	$::= (\langle expr \rangle)\langle op \rangle(\langle expr \rangle)$	(1.a)
	$\mid \langle coef \rangle \times \langle var \rangle$	(1.b)
$\langle op \rangle$	$::= \div \mid \times \mid + \mid -$	(2.a), (2.b), (2.c), (2.d)
$\langle coef \rangle$	$::= C1 \mid C2$	(3.a), (3.b)
$\langle var \rangle$	$::= V1 \mid V2$	(4.a), (4.b)

Table 3.2 Production of a terminal element

MOD	Rule	Current element state
166 MOD 2	0	$(<expr>) <op>(<expr>)$
225 MOD 2	1	$(<coef> \times <var>)<op>(<expr>)$
180 MOD 2	0	$(c_1 \times <var>)<op>(<expr>)$
132 MOD 2	0	$(c_1 \times v_1) <op>(<expr>)$
187 MOD 4	3	$(c_1 \times v_1) \div (<expr>)$
219 MOD 2	1	$(c_1 \times v_1) \div (<coef> \times <var>)$
179 MOD 2	1	$(c_1 \times v_1) \div (c_2 \times <var>)$
249 MOD 2	1	$(c_1 \times v_1) \div (c_2 \times v_2)$

is the MOD rule defined as (codon integer value) MOD (number of rules for the current non-terminal), where MOD is the modulus operator (%).

Now it is shown how the example chromosome above generates a symbolic feature expression using the grammar provided in Table 3.1, which is later evaluated as a numerical feature. Using the start symbol $\mathcal{S} = <expr>$, there are 2 production rules to choose from, (1.a) and (1.b). The MOD operation on the current codon becomes (166)MOD(2) which yields 0; hence, rule (1.a) is chosen. The successive application of rules in Table 3.2 shows how a feature expression is generated by the example chromosome.

The derivation sequence above can be maintained as a tree. The derivation tree is then reduced to the standard GP syntax tree as in Fig. 3.2.

Fig. 3.2 Feature generation tree

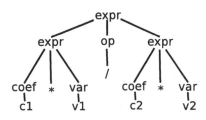

GE uses standard concepts in canonical GAs to evolve the chromosomes, thereby generating new derivation sequences, new feature expressions and new numerical features. For a given chromosome, the resultant feature can be a terminal ($\mathscr{T}$) or a non-terminal ($\mathscr{N}$) element. If the chromosome runs out of codons without producing a terminal element as above, it is wrapped around and the codons are reused from the beginning. This could lead to incomplete chromosomes if the mapping never produces a terminal element, which is addressed by introducing a limit on the allowed chromosome wrappings and returning a poor fitness score. An in-depth explanation on the usage of GE can be found in the original GE paper by O'Neill in [1].

3.5 gramEvol Package

This section introduces the R package gramEvol [10], which facilitates the construction and execution of programs in R [11] using GE. With thousands of third-party packages, covering different areas of pure and applied statistics and probability, R has evolved to become the de facto language for data sciences and machine learning. Appealing to this community, we present our implementations in R. The design goal is to evolve programs in R from a concise description. While it is possible to call existing GE libraries from R, a native implementation was developed since it has the following advantages:

1. Only R's base package is used for evolutionary operations and grammar processing as well as parsing and running generated programs, eliminating the need for third-party libraries and external dependencies.
2. R's data structures (specifically list) are used to define a BNF grammar, removing a BNF interpenetration or compilation step and allowing dynamic grammar manipulation and rapid prototyping.

One disadvantage of gramEvol is its speed compared with compiled C++ GE libraries, specifically libGE [12] and AGE [13]. It is assumed that the computational overhead of processing the fitness function is greater than the overhead of GE operators, and hence forms the major bottleneck. An R implementation can thus be accelerated by moving the fitness function computation to a compiled language such as C, C++ or Fortran, as is common practice and facilitated by packages such as Rcpp [14]. The usage of gramEvol package is demonstrated by attempting a symbolic regression.

Symbolic regression using GE Symbolic regression is the process of discovering a function, in symbolic form, which fits a given set of data. Evolutionary algorithms such as GP and GE can be applied to this task.

Rediscovering scientific laws has been studied using different heuristic techniques [15] including, GP [16] and gene expression programming [17]. Here, as an example of symbolic regression, we rediscover Kepler's Law in which the goal is to find a relationship between orbital periods and the distances of planets from the sun, given the normalised data in Table 3.3.

Table 3.3 Orbit period and distance from the sun for planets in solar system

Planet	Distance	Period
Venus	0.72	0.61
Earth	1.00	1.00
Mars	1.52	1.84
Jupiter	5.20	11.90
Saturn	9.53	29.40
Uranus	19.10	83.50

To use grammatical evolution to find this relationship from the data, we define a grammar as illustrated in Table 3.4.

The first step for using `gramEvol` is loading the grammar defined in Table 3.4:

```
R> library(gramEvol)
R> ruleDef <-
+ list(list("expr", list("<expr><op><expr>", "<subexpr>")),
+ list("sub-expr", list("<func>(<expr>)", <var>",
+                                          "<var>^<n>")),
+ list("func", list("sqrt", "log", "sin", "cos")),
+ list("op", list("+", "-", "*")),
+ list("var", list("distance", "<n>")),
+ list("n", list("1", "2", "3", "4")))

R> grammarDef <- CreateGrammar(ruleDef, startSymb
+                                          = "<expr>")
```

We use the following equation to normalise the error, adjusting the impact of error on small values (e.g. Venus) versus large values (e.g. Uranus):

$$e = \frac{1}{N} \sum \log(1 + |p - \hat{p}|) \tag{3.1}$$

Table 3.4 Grammar for discovering Kepler's equation

$\mathcal{N} = \{expr, sub\text{-}expr, func, op, var, n\}$
$\mathcal{T} = \{+, -, \times, \hat{}, \log, sqrt, sin, cos, \text{distance}, 1, 2, 3, 4, (,)\}$
$\mathcal{S} = <expr>$

Production rules: $\mathcal{R}$

$\langle expr \rangle$	$::= \langle expr \rangle \langle op \rangle \langle expr \rangle \mid \langle sub\text{-}expr \rangle$	(1.a), (1.b)
$\langle sub\text{-}expr \rangle$	$::= \langle func \rangle(\langle var \rangle) \mid \langle var \rangle \mid \langle var \rangle \hat{}\ \langle n \rangle$	(2.a), (2.b), (2.c), (2.d)
$\langle func \rangle$	$::= \log \mid sqrt \mid sin \mid cos$	(3.a), (3.b), (3.c), (3.d)
$\langle op \rangle$	$::= + \mid - \mid \times$	(4.a), (4.b), (4.c)
$\langle var \rangle$	$::= \text{distance} \mid \langle n \rangle$	(5.a), (5.b)
$\langle n \rangle$	$::= 1 \mid 2 \mid 3 \mid 4$	(6.a), (6.b), (6.c), (6.d)

where e is the normalised error, N is the number of samples, p is the orbital period and $\hat{p}$ is the result of symbolical regression. We implement this as the fitness function SymRegFitFunc:

```
R> planets <- c("Venus", "Earth", "Mars", "Jupiter",
                                    "Saturn", "Uranus")
R> distance <- c(0.72, 1.00, 1.52, 5.20, 9.53, 19.10)
R> period <- c(0.61, 1.00, 1.84, 11.90, 29.40, 83.50)

R> SymRegFitFunc <- function(expr) {
+    result <- EvalExpressions(expr)
+    if (any(is.nan(result)))
+       return(Inf)
+    return (mean(log(1 + abs(period - result))))
+ }
```

Notice that a very low fitness (infinite error) is assigned to invalid expressions in the fitness function.

As GE is a stochastic optimiser, every run may result in a slightly different answer. For the purpose of reproducibility, the random generator seed value is set to a constant value:

```
R> set.seed(100)
```

GrammaticalEvolution is now ready to run. In this example, the number of generations is limited to 100 and the population size is set to 50 by the default values of GrammaticalEvolution. As the best possible outcome and its error are known, terminationFitness is computed and set to terminate GE when the Kepler's equation is found.

```
R> ge <- GrammaticalEvolution(grammarDef, SymRegFitFunc,
+                             terminationFitness = 0.021)

R> ge$bestExpression

[1] "sqrt(distance^3)"

R> data.frame(distance, period, Kepler = sqrt(distance^3),
+             GE = EvalExpressions(ge$bestExpression))

  distance period     Kepler          GE
1     0.72   0.61  0.6109403   0.6109403
2     1.00   1.00  1.0000000   1.0000000
3     1.52   1.84  1.8739819   1.8739819
4     5.20  11.90 11.8578244  11.8578244
5     9.53  29.40 29.4197753  29.4197753
6    19.10  83.50 83.4737743  83.4737743
```

Table 3.5 Summary of grammatical evolution's performance for 100 runs of symbolic regression example

Value	Minimum	Average	Maximum
Error	0.00	0.92	1.61
No. of generations	2.00	77.46	100.00
Time (s)	0.40	4.16	20.31

Other GE runs find expressions such as `sqrt(distance)*distance` or `sqrt(distance^3+cos(distance)*log(1^4))` which all simplify to Kepler's Third Law which states:

$$period^2 = constant \times distance^3 \qquad (3.2)$$

The fitness function handles invalid values (e.g. log(−1)) by assigning a low fitness (infinite error) to any chromosome with an invalid value. However, R may show warnings about NaNs being produced. To suppress these warnings, it is enough to wrap `EvalExpressions` in fitness function inside `suppressWarnings`:

```
R> result <- suppressWarnings(EvalExpressions(expr))
```

As an incomplete search is performed, sometimes the GE fails to find a perfect solution. In such cases, a symbolic result with error is presented (i.e. log(*distance*) in `sqrt(distance^3+log(distance))`). To characterise this behaviour, the code was run 100 times and its error from Kepler equation was noted. The measurements were performed on a single thread on a 3.40 GHz Intel Core i7-2600 CPU. To ensure reproducibility, `set.seed(0)` was executed before running the code. The results are presented in Table 3.5. Notice that the average performance can be improved at the expense of time, by increasing the GE's number of generation `iterations` or population size `popSize`.

`GrammaticalEvolution` allows monitoring the status of each generation using a callback function. This function, if provided to parameter `monitorFunc`, receives an object similar to the return value of `GrammaticalEvolution`. For example, the following function prints the current generation, the best chromosome's expression and its error and produces a plot:

```
R> customMonitorFunc <- function(results){
+    cat("-------------------\n")
+    cat("Current Generation: ",
+                results$gaSummary$currentIteration, "\n")
+    cat("Best fitness:",
+                min(results$gaSummary$evaluations), "\n")
+    cat("Best Expression:", results$bestExpression, "\n")
+    vals <- EvalExpressions(results$bestExpression)
```

```
+   plot(distance, period, type = "o", log = "y")
+   lines(distance, vals, type = "o", col = "red")
+   legend("bottomright",
+     legend = c("Observed", "GE fit"),
+     col = c("black", "red"),
+     pch = 1,
+     lty = 1)
+ }

R> ge <- GrammaticalEvolution(grammarDef, SymRegFitFunc,
+       terminationFitness = 0.021,
+       monitorFunc=customMonitorFunc)

R> print(ge$bestExpression)
R> print(data.frame(distance, period,
+       Kepler=sqrt(distance^3),
+       GE=EvalExpressions(ge$bestExpression)))
```

References

1. M. O'Neill, C. Ryan, Grammatical evolution. IEEE Trans. Evol. Comput. **5**(4), 349–358 (2001)
2. Il.-S. Oh, J.-S. Lee, B.-R. Moon, Hybrid genetic algorithms for feature selection. IEEE Trans. Pattern Anal. Mach. Intell. **26**(11), 1424–1437 (2004)
3. J. Yang, V. Honavar, Feature subset selection using a genetic algorithm. IEEE Intell. Syst. Appl. **13**(2), 44–49 (1998)
4. C.-L. Huang, C.-J. Wang, A ga-based feature selection and parameters optimization for support vector machines. Expert Syst. Appl. **31**(2), 231–240 (2006)
5. D.P. Muni, N.R. Pal, J. Das, Genetic programming for simultaneous feature selection and classifier design. IEEE Trans. Syst. Man Cybern. Part B Cybern. **36**(1), 106–117 (2006)
6. J.-Y. Lin, H.-R. Ke, B.-C. Chien, W.-P. Yang, Classifier design with feature selection and feature extraction using layered genetic programming. Expert Syst. Appl. **34**(2), 1384–1393 (2008)
7. M. Srinivas, L.M. Patnaik, Genetic algorithms: a survey. Computer **27**(6), 17–26 (1994)
8. M. Sipser, in *Context-Free Grammars*. Introduction to the Theory of Computation, chap. 2 (PWS Publishing, Boston, 1997), pp. 91–122
9. D.E. Knuth, Backus normal form vs Backus Naur form. Commun. ACM **7**(12), 735–736 (1964)
10. A.M. de Silva, F. Noorian, R package for grammatical evolution (2013), http://cran.r-project.org/web/packages/gramEvol/index.html. Accessed Nov 2013
11. R Foundation for Statistical Computing, *R: a Language and Environment for Statistical Computing*, http://www.r-project.org/
12. M. Nicolau, libGE C++ Library (2006), http://bds.ul.ie/libGE/. Stable release 0.26
13. A. Nohejl, AGE: algorithms for grammar-based evolution (2011), http://nohejl.name/age/. Version 1.1.1
14. D. Eddelbuettel, R. Francois, Rcpp: seamless r and c++ integration. J. Stat. Softw. **40**(8), 1–18 (2011)
15. P. Langley, H.A. Simon, G.L. Bradshaw, in *Heuristics for Empirical Discovery*. Computational Models of Learning (Springer, Berlin, 1987), pp. 21–54
16. J.R. Koza, *Genetic Programming: On the Programming of Computers by Means of Natural Selection*, vol. 1 (MIT Press, Washington, 1992)
17. P. Sherrod, DTReg: gene expression programming (2014), http://www.dtreg.com/gep.htm

Chapter 4
Grammar Based Feature Generation

Abstract This chapter describes the core functionality expected from a rule based feature generation framework. The mechanism in which the rule sequences can be invoked to generate string feature expressions is illustrated for simple and complex feature expressions, i.e. technical indicators. The issue of selecting features from large feature spaces and selective feature pruning strategies that can be used to contain the most informative features is also presented. The importance of feature selection in a feature generation framework is highlighted.

Keywords Context-free grammars · Feature generation framework · Wavelet based time-series prediction · Technical indicators · Heuristic feature pruning · Hybrid feature selection and generation

4.1 Automatic Feature Generation

Instead of manually constructing feature vectors, the methodology proposed in this chapter is to automatically generate a large pool of potential features, and employ feature selection to find a suitable subset. Although the proper selection of analytical features is of utmost importance, it is often overlooked and it is doubtful if the best possible models can be found via manual means. Context-free grammars (CFGs) are employed as a systematic way of generating suitable features. The notion of grammar families as a compact representation to generate a broad class of features is explained. Special attention is given to feature pruning and implementation issues arising as a result of processing a large number of features. Furthermore, grammatical evolution (GE) is proposed as a convenient technique to combine feature generation and feature selection (FS) without selective feature pruning.

4.2 Feature Generation Using Context-Free Grammars

Automated feature generation is the process of generating numerical descriptions of some data instances. A manual approach involves identifying certain characteristics of the time-series under consideration, and then deriving numerical measures to

A.M. De Silva and P.H.W. Leong, *Grammar-Based Feature Generation*
for Time-Series Prediction, SpringerBriefs in Computational Intelligence,
DOI 10.1007/978-981-287-411-5_4

describe them [1]. As suggested in Chap. 1, one aim of the feature generation system is to enable experts to provide guidelines to the automated feature generator but for the computer to do all the tedious work of generating and selecting appropriate feature subsets for the prediction task at hand.

4.2.1 Grammar Based Feature Generation Systems

Automated feature generation is an interesting research area since it may systematically produce and select better feature combinations than those designed by humans. A general framework for function-based feature generation using context-free grammars was first proposed by Markovitch and Rosenstein [2]. Such grammars are used in linguistics to describe sentence structure and words of a natural language and in computer science to describe the structure of programming languages [3]. Markovitch and Rosenstein generated features strongly related to the target using decision trees. Unfortunately, the technique is only suitable for problems where the features are apparent from the problem definition. Eads et al. [1] and Pachet and Roy [4] addressed supervised time-series classification using standard genetic programming to discover a set of fundamental signal processing operations via a grammatical structure. Both these works conclude that conventional classifiers trained using raw data as features can be improved by training the same classifiers with grammar generated features. Standard genetic programming was also used by Ritthof et al. [5] to combine feature generation and feature selection, and applied to the interpretation of chromatography time-series. Reference [5] used arithmetic operators in their grammar to expand the feature space while Eads et al. extracted time-series information using operators such as the mean, delay, derivative, integral, etc. Reference [6] used genetic programming with a set of mathematical transformation operators, e.g. sin, cos, $+$, $-$, sqrt, etc. to produce features of the raw vibration signals from a rotating machine (fault classification). The same framework was applied to the diagnosis of breast cancer [7]. Both works reported improved classification accuracies. Reference [8] proposed a feature generation algorithm for analysing splice-site reduction in biology, and achieved an improvement of approximately 6 % compared with using well-known features. Reference [9] applied a co-evolutionary feature generation approach where multiple populations were evolved simultaneously. Primary operators applicable to images, e.g. different filters, image norms, scalar operators etc. were defined and the best operator sequences (or processing steps embedded in chromosomes) were used for synthetic aperture radar (SAR) image recognition. The approach proved to be robust under different operating conditions.

The framework illustrated in this chapter uses a range of such operators to expand the feature space. Parameters of operators are selected to maximize the prediction performance of the ML algorithm under consideration. There has been little previous work on feature selection techniques within feature generation frameworks with the aim of selecting robust features. This problem is extremely important in predicting non-stationary time-series as emphasized in Chaps. 2 and 3. Comparisons

to well-known features are also missing in the literature. Reference [10] used GE
to select features for detecting epileptic oscillations within clinical intracranial elec-
troencephalogram (iEEG) recordings of patients with epilepsy. GE has also been
used in computational finance, credit rating and corporate failure prediction, music
and robot control applications (see [11] for a survey). None of these works however
formulate a well organized general framework that is extensible and customizable.
Furthermore, none of the works have applied their techniques to financial time-series
prediction using technical indicators or electricity load time-series prediction or foreign exchange client trade volume time-series predication.

4.2.2 Feature Generation Framework Architecture

This work uses CFGs to systematically guide the hierarchical generation of a pool
of candidate features. Unlike genetic algorithms in feature construction [6], a CFG
framework facilitates visualization of the feature generation process, aiding in the
monitoring of the features generated. The layered organization used to generate
features is shown in Table 4.1. At time k, the base layer consists of the observed variables $x_k^{(1)}, \ldots, x_k^{(m)}$ and the derived variables $f_k^{(1)}, \ldots, f_k^{(l)}$. Using the notation in
Table 4.4, O, H, L, C are the observed variables and M, U, D are the derived variables
in the financial time-series prediction context.

The transformation operators in the combinatorial and transformation layer consists of base operators and running operators. Examples of base operators are the
first difference and absolute value. Running operators use sliding windows of length
n to identify local features. By varying n, long or short history information can be
captured systematically. The combinatorial operators fuse information across variables to produce more features which are passed to the user defined layers, and are
defined in Table 4.2. Domain transformations (e.g. wavelets and the discrete Fourier

Table 4.1 Layered organization of operators for feature generation

User defined layers		e.g. EMA($c_k^{(i)}$, n)
Combinatorial and transformation layer expressions $c_k^{(i)}$ are passed to the higher layers		
Combinatorial and transformation layer	Fractional combinations	e.g. $(b_k^{(i)} - b_k^{(j)})/b_k^{(j)}$
	Additive combinations	e.g. $b_k^{(i)} \pm b_k^{(j)}$
	Base operators	e.g. $\log(b_k^{(i)})$
	Running operators	e.g. func($b_k^{(i)}$) (see Table 4.2)
Base layer elements $b_k^{(i)}$ are passed to the transformation layer		
Base layer	Derived variables	$f_k^{(1)}, \ldots, f_k^{(l)}$
	Observed variables	$x_k^{(1)}, \ldots, x_k^{(m)}$

Table 4.2 Base operators and running operators

Base operators		Running operators (window size n)	
diff(x, n)	$x_k - x_{k-n}$	sma(x)	Simple moving average
log(x)	Natural log	wilder(x)	Wilder exponential moving average
delt(x)	$(x_k - x_{k-1})/x_k$	ema(x)	Exponential moving average
abs(x)	$\lvert x_k \rvert$	wma(x)	Weighted moving average
lag(x, n)	x_{k-n}	max(x)	Maximum value
sin(x)		min(x)	Minimum value
cos(x)		sd(x)	Standard deviation
...		sum(x)	Summation
		meandev(x)	Mean deviation
		skewness(x)	Skewness
		kurtosis(x)	Kurtosis
		median(x)	Median
		histwin(x)	A history window (a lag sequence)

transform) are also useful in constructing informative features. For example, the wavelet transformation has been used for multi-resolution analysis of stock data, and are able to capture information on different time-scales that are not obvious from the original time-series [12]. Such transformed variables can easily be incorporated to the framework through their inclusion in the base layer as derived variables. Additionally, transformations can be defined in a higher layer, e.g. in the combinatorial and transformation layers.

The operators defined in Table 4.2 can be used to generate a broad class of features. In the next section, we describe how the framework can generate features from a grammar based on wavelet components for electricity load time-series prediction.

4.3 Wavelet Based Grammar for Electricity Load Time-Series Prediction

In order to demonstrate how features are generated, the Backus-Naur form (BNF) grammar in Table 4.3 is used to find features relevant to the prediction of peak electricity load time-series. The operator notation is as in Table 4.2 and E is the half-hourly electricity load time-series. D^1, D^2, D^3 and $S = S^3$ are the 3-level wavelet decomposition components of E. Wavelet transformation background theory is deferred to Appendix A.

Recall that a CFG can be described by $(\mathcal{T}, \mathcal{N}, \mathcal{R}, \mathcal{S})$ as explained in Sect. 3.3. The production rules $\mathcal{R}$ for this grammar are organized into 5 groups. Group 1 has 3 rules (1.a)–(1.c), group 2 has 3 rules (2.a)–(2.c) and so on. The total number of production rules is 17, (1.a)–(5.e) ($\lvert R \rvert = 17$). Each production rule has a

Table 4.3 Wavelet based grammar for peak electricity load time-series prediction

$\mathcal{T} = \{\texttt{abs}, \texttt{delt}, \texttt{diff}, \texttt{lag}, \texttt{sma}, \texttt{sd}, \texttt{meandev}, \texttt{histwin}, \texttt{E}, \texttt{D}^1, \texttt{D}^2, \texttt{D}^3, \texttt{S}, \texttt{n}, \texttt{k}, \texttt{(,)}\}$

$\mathcal{N} = \{expr, base\text{-}var, pre\text{-}op, base\text{-}op, var\}$

$\mathcal{S} = \langle expr \rangle$

$\mathcal{R}$ Production rules

$\langle expr \rangle$	::=	$\texttt{histwin}(\langle base\text{-}var \rangle, \texttt{n})$	(1.a)
		$\mid \langle pre\text{-}op \rangle(\langle base\text{-}var \rangle, \texttt{n})$	(1.b)
		$\mid \langle base\text{-}var \rangle$	(1.c)
$\langle base\text{-}var \rangle$	::=	$\langle base\text{-}op \rangle(\langle var \rangle)$	(2.a)
		$\mid \texttt{lag}(\langle var \rangle, \texttt{k})$	(2.b)
		$\mid \langle pre\text{-}op \rangle(\langle var \rangle, \texttt{n})$	(2.c)
$\langle pre\text{-}op \rangle$	::=	$\texttt{sma} \mid \texttt{sd} \mid \texttt{meandev}$	(3.a), (3.b), (3.c)
$\langle base\text{-}op \rangle$	::=	$\texttt{delt} \mid \texttt{diff} \mid \texttt{abs}$	(4.a), (4.b), (4.c)
$\langle var \rangle$	::=	$\texttt{E} \mid \texttt{D}^1 \mid \texttt{D}^2 \mid \texttt{D}^3 \mid \texttt{S}$	(5.a), (5.b), (5.c), (5.d), (5.e)

head (left hand side), a non-terminal symbol in $\mathcal{N}$, that is assigned by the string of symbols in the body (right hand side). Multiple production rules in the same group are delimited by the pipe "|". There are 5 non-terminal symbols ($\mathcal{N}$) which are denoted in the production rules by <·>. A feature generated by each rule corresponds to a particular layer in Table 4.1. Rule (2.b) adds different lags of observed (E) and derived variables ($\texttt{D}^1, \texttt{D}^2, \texttt{D}^3$ and S) to the generated feature pool. Rule group 2 can be considered as producing features in the combinatorial and transformational layer. Rule (1.a) produces user defined layer features by applying different operators on variable combinations. The rules (1.b) and (1.c) act as mere linking rules that invoke other rules.

As described in Sect. 3.3, to generate a specific feature, the feature generation sequence is initiated with the start symbol $\mathcal{S} = $ <*expr*>. The production rules are sequentially invoked on the left-most non-terminal to generate features. The generation of a simple feature is illustrated first.

The production rule sequence (1.b)→(3.a)→(2.b)→(5.a) generates sma(lag (E, k),n) which is a running operator in the combinatorial and transformation layer, e.g. for n = 10 and k = 0, the feature can be interpreted as the simple moving average of peak load for 10 days. This rule invoking sequence is compactly represented in Fig. 4.1. The circles represent the leftmost non-terminals on which the rules are invoked and the rule number is indicated below the arrows.

Fig. 4.1 Step-wise generation of exponential moving average as a feature from the wavelet grammar in Table 4.3

This can also be illustrated by the following notation.

Invoking rule (1.b): $<expr> ::= <pre-op>(<base-var>, n)$

Invoking rule (3.a): $(<pre-op>) ::= \text{sma}$

Invoking rule (2.b): $<base-var> ::= \text{lag}(<var>, k)$

Invoking rule (5.a): $<var> ::= \text{E}$

Rule sequences like above generate primary features such as moving averages. More complex feature generation sequences, which produce technical indicator type formulae are demonstrated later in this chapter.

4.4 Grammar Families

By defining well organized compact grammar families (instead of a single grammar), the number of generated features can be significantly reduced. More importantly, the generated features can now be more informative and interpretable because they are systematically organized by a human expert. Once all the features are generated, feature selection and dimensionality reduction can be applied to select the best features.

This is illustrated by introducing the grammar families designed to generate *technical indicators*. If a single grammar was designed to generate all the technical indicators surveyed, the number of features will be extremely large. Grammar families were designed to generate a superset of the standard indicators, with an aim of identifying other potentially new technical indicators.

4.4.1 Technical Indicators

Technical indicators are formulae that identify patterns and market trends in financial markets [13] which are developed from models for price and volume. The seminal work of [14] on technical analysis still remains in use to the present. Technical indicators can be broadly classified as trend, momentum, volatility and volume indicators. A trend analysis studies price charts using the moving average filters. The moving average filter gives smooth price estimates and identifies overall trend patterns. The exponential moving average (EMA), simple moving average (SMA) and weighted moving average (WMA) filters are some of the standard trend indicators. Momentum measures the variation of price in a given time. Momentum indicators identify overbought and oversold positions and start of new trends. Rate of convergence (ROC), relative strength index (RSI) and average directional index (ADX) are some commonly used momentum indicators. Volatility indicators like Bollinger bands (BB) identify the uncertainty in the market via statistical variance of price movements. Volume indicators identify the volumes of trade that have the potential to cause market movements, and the money flow index (MFI) is one such example. Table 4.5 summarizes some well-known and standard trend, momentum, volatility and volume indicators and its notation is in Table 4.4.

Table 4.4 Symbol notation. The subscript k denotes the current day

Symbol	Interpretation
O	Opening price of the current day
H	Highest price of the day
L	Lowest price of the day
C	Closing price of the day
V	Traded volume for the day
M	Typical (average) price, $(H_k + L_k + C_k)/3$
U	Upward price change, $\max(0, C_k - C_{k-1})$
D	Downward price change, $\min(0, C_k - C_{k-1})$
F	Money flow at a given time, $M \times V$
F^+	Positive money flow, $\max(0, F_k - F_{k-1})$
F^-	Negative money flow, $\min(0, F_k - F_{k-1})$
H^+	Highest high price of the day, $\max(H_{k-i})_{i=0}^{n-1}$
L^-	Lowest low price of the day, $\min(L_{k-i})_{i=0}^{n-1}$
i^+	Days elapsed since the last highest price, $\arg\max_i P_{k-i}, i = 0, 1, \ldots, n-1$
i^-	Days elapsed since the last lowest price, $\arg\min_i P_{k-i}, i = 0, 1, \ldots, n-1$

4.4.2 Generating Technical Indicators Using Grammar Families

This section demonstrates how a set of grammar families generate a broad class of features with the technical indicators summarized in Table 4.5 as particular cases. seven grammar families generate all the technical indicators in Table 4.5 as particular cases. The CFG based framework is flexible in that, (i) the number of grammar families and the organization of the production rules can be adapted (ii) the user is able to design a sufficiently large grammar to capture as much information as possible with a manageable feature space and (iii) the user can incorporate domain knowledge by choosing appropriate derived variables and production rules. The grammar families 1 and 2 are brought in Tables 4.6 and 4.7 and the rest in Appendix B Tables B.1–B.5.

The grammar family 1 is used to illustrate the generation of standard technical indicators A/D oscillator, CLV, bias and ROC. Figure 4.2 shows the production rules invoked to generate the technical indicator A/D oscillator and the intermediate states produced in the process. The start symbol for grammar family 1 is *<L3>*. By invoking rule (1.b) on *<L3>*, the intermediate non-terminal expression (*<L2>*) ÷ (*<L2>*) is produced. Since there are 2 individual non-terminal elements in this intermediate expression, the leftmost non-terminal is always chosen (hence circled in figure). Figure 4.2 can be understood in this manner and the resultant expression can be verified to be the A/D oscillator formula provided in Table 4.5 with C lagged by

Table 4.5 Standard trend, volatility and volume indicators

Indicator name	Acronym	Formula	Parameters
Simple moving average	SMA	$\frac{1}{n}\sum_{i=0}^{n-1} P_{k-i}$	$n = 5, 15, 30$
Weighted moving average	WMA	$\frac{1}{n}\sum_{i=0}^{n-1}(n-i)P_{k-i}$	$n = 15, 5, 30$
Exponential moving average	EMA	$\sum_{i=0}^{n-1}\alpha(1-\alpha)^i P_{k-i}$	$n = 5, 15, 30$ $\alpha = 2/(n+1)$
Disparity	DIS	$P_k/\mathrm{ema}(P_k, n)$	$n = 5, 10$
Bias	BIAS	$(P_k - \mathrm{sma}(P_k, n))/n$	$n = 5, 10$
Bollinger bands	BB	$\mathrm{sma}(P_k, n) \pm 2\sigma$	$n = 15$
Chaikin volatility	–	$\mathrm{sma}(H_k - L_k, n) \pm 2\sigma$	$n = 15$
Average true range	ATR	$\max(H_k - L_k, \lvert H_k - C_{k-n}\rvert, \lvert L_k - C_{k-1}\rvert)$	–
Money flow index	MFI	$(1 + R)/R$	$n = 14$
Aroon indicator (Up, Down)	–	$(n - i^+)/n, (n - i^-)/n$	$n = 6, 12, 24$
Rate of convergence	ROC	$(C_k - C_{k-n})/C_{k-n}$	$n = 1$
Commodity channel index	CCI	$(M_k - \mathrm{sma}(M_k, n))/0.015\bar{\sigma}$	$n = 15$
Relative strength index	RSI	$RS/(1 + RS)$	$n = 14$
Moving average convergence divergence	MACD	$\mathrm{ema}(C_k, n_1) - \mathrm{ema}(C_k, n_2), n_1 > n_2$	$n_1 = 26, n_2 = 12$
Momentum	MOM	$C_k - C_{k-n}$	$n = 4$
William's indicator	R	$(H_{k-n}^+ - C_k)/(H_{k-n}^+ - L_{k-n}^-)$	$n = 15$
Stochastic oscillator	K	$(C_{k-n} - L_{k-n}^-)/(H_{k-n}^+ - L_{k-n}^-)$	$n = 14$
Stochastic indicator	D	$\mathrm{sma}(K(n_1), n_2)$	$n_1 = 15, n_2 = 5$
Slow stochastic indicator	Slow D	$\mathrm{sma}(D(n_1, n_2), n_3)$	$n_1 = 15, n_2, n_3 = 3$
Close location value	CLV	$((C_k - L_k) - (H_k - C_k))/(H_k - L_k)$	–
Price Oscillator	OSCP	$(\mathrm{sma}(P_k, n_1) - \mathrm{sma}(P_k, n_2))/\mathrm{sma}(P_k, n_1)$	$n_1 = 5, n_2 = 10$
Accumulation/Distribution Oscillator	ADO	$(H_k - C_{k-1})/(H_k - L_k)$	–

Additional notation Money ratio $R = \Sigma_0^{n-1}F_i^+/F_i^-$. $RS = \mathrm{ema}(U, n)/\mathrm{ema}(D, n)$

$k = 0$ and L lagged by $k = 1$, i.e. $(H - \mathrm{lag}(C, 1)) \div (H - \mathrm{lag}(L, 0))$. The generation of CLV and ROC by grammar family 1 is deferred to Appendix B.

In a similar fashion, Fig. 4.3 illustrates how the technical indicator disparity is generated using the family 2. The final expression, $(\mathrm{lag}(C, k)) \div (\mathrm{ema}(\mathrm{lag}(C, k), n))$ provides the disparity with C lagged by $k = 0$ as can be verified with the formula in Table 4.5. All 7 grammar families were defined to generate one or more

Table 4.6 Grammar family 1

Family 1		
$\mathcal{N} = \{L1, L2, L3\}$		
$\mathcal{T} = \{-, \div, \text{lag}, \text{sma}, \text{meandev}, \text{sum}, H, L, C, M, n, k, N, (,) \}$		
$\mathcal{S} = \{L3\}$		
Production rules: $\mathcal{R}$		

$\langle L3 \rangle$	::= $(\langle L2 \rangle) \div (\text{lag}(\langle L2 \rangle, k)) \mid (\langle L2 \rangle) \div (\langle L2 \rangle)$	(1.a), (1.b)
	$\mid (((\langle L2 \rangle)) - ((\langle L2 \rangle))) \div N \mid \langle L2 \rangle$	(1.c), (1.d)
$\langle L2 \rangle$	::= $\langle L1 \rangle - \text{lag}(\langle L1 \rangle, k) \mid \langle L1 \rangle - \text{sma}(\langle L1 \rangle, n)$	(2.a), (2.b)
	$\mid \text{meandev}(\langle L1 \rangle, n) \mid \text{sum}(\langle L1 \rangle, n) \mid \langle L1 \rangle$	(2.c), (2.d), (2.e)
$\langle L1 \rangle$	::= $H \mid L \mid C \mid M$	(3.a), (3.b), (3.c), (3.d)

Table 4.7 Grammar family 2

Family 2		
$\mathcal{N} = \{L1, L2, L3, L4\}$		
$\mathcal{T} = \{-, \div, \text{lag}, \text{sma}, \text{ema}, \text{wma}, H, L, C, M, \text{delt}, \text{diff}, n, k, (,) \}$		
$\mathcal{S} = \{L4\}$		
Production rules: $\mathcal{R}$		

$\langle L4 \rangle$	::= $(\langle L3 \rangle) \div (\langle L3 \rangle) \mid (\langle L3 \rangle - \langle L3 \rangle) \mid \langle L3 \rangle$	(1.a), (1.b), (1.c)
$\langle L3 \rangle$	::= $\text{ema}(\langle L2 \rangle, n) \mid \text{sma}(\langle L2 \rangle, n) \mid \text{wma}(\langle L2 \rangle, n)$	(2.a), (2.b), (2.c)
	$\mid \text{sma}(\text{ema}(\langle L2 \rangle, n), n) \mid \langle L2 \rangle$	(2.d), (2.e)
$\langle L2 \rangle$	::= $\text{diff}(\langle L1 \rangle) \mid \text{delt}(\langle L1 \rangle) \mid \text{lag}(\langle L1 \rangle, k)$	(3.a), (3.b), (3.c)
$\langle L1 \rangle$	::= $H \mid L \mid C \mid M$	(4.a), (4.b), (4.c), (4.d)

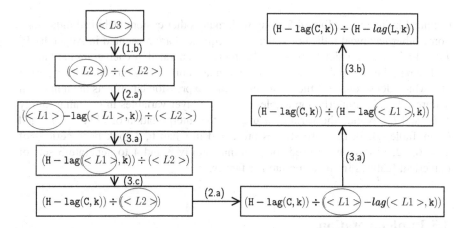

Fig. 4.2 Step-wise generation of ADO indicator. The *circles* indicate the current non-terminals on which the rules are invoked and the rule number is indicated

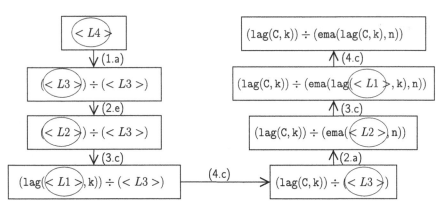

Fig. 4.3 Step-wise generation of disparity indicator. The *circles* indicate the current non-terminals on which the rules are invoked and the rule number is indicated

Table 4.8 Standard technical indicators generated by each grammar family

Family no	Standard technical indicators generated
1	CLV, CCI, ROC, ADO, Bias, Lagged prices
2	EMA, SMA, WMA, Lagged prices, Disparity, MACD, SD, OSCP
3	R, K, D, Slow D
4	Aroon
5	RSI, MFI
6	BB, Chakin volatility
7	Volume related indicators

technical indicators in Table 4.5 along with many other custom technical indicators. Some other technical indicator generation sequences are presented in Appendix B. Table 4.8 indicates which technical indicators are generated by each grammar family.

The base layer variables H^+, L^- in grammar family 3 rely on a look-back period of n which does not affect the symbolic feature expression but affects the numerical features. This is similarly applicable to the base layer variables in grammar family 4, i^+ and i^-. Notice the variable n used in the interpretation of H^+, L^-, i^+ and i^- in Table 4.4. In the case studies presented in Chap. 6, 3 look back-periods of $n = 6, 12, 24$ days were used for grammar family 3 and 4 to generate 3 sets of numerical features from each grammar family.

4.5 Implementation

This section discusses the practical aspects of the feature generation framework and tactics that can be adopted in the implementation.

4.5.1 Pruning Strategies

Some of the generated features are parametrized by the window-size n and lag k (The parameter values used were $n = 5, 15, 30$ and $k = 0, 1, 2, 3, 4, 5, 6$ in our case studies in Chap. 6). This means that a feature expression can produce multiple numerical features, e.g. the feature expression EMA(H,n) generates 3 numerical features EMA(H,5), EMA(H,15) and EMA(H,30). Clearly, the language size increases dramatically if parameters n and k have a wide range. Appropriate feature pruning can be used to retain only the top ranking features. An alternative approach is to generate a very large number of features and then choose a subset of the features. Unfortunately, the larger the search space, the harder it becomes to mine for better features. A computationally more efficient approach is to limit the number of features generated in the first place. This can be achieved by introducing pruning mechanisms to ensure the number of features remain within a manageable range. In the following, the primary pruning strategies are discussed.

(i) Feature pruning can be implicitly achieved by carefully designing the grammar structure. The first step towards this was to introduce grammar families in Sect. 4.4. By defining well organized compact grammar families (instead of a single grammar), the number of generated features can be significantly reduced.

(ii) Limiting the number of production rules in each rule group can also avoid the feature space becoming too large. Constructing focused grammar families for different groups of technical indicators was used to achieve this.

(iii) Grammar family 4 in Table B.2 has separate production rules for <L1> and <L2> for the numerator and denominator terms. This reduces the number of permutations in comparison to using a single rule to generate fractional features.

(iv) Avoiding invalid combination of terms in a production rule avoids generation of meaningless features, e.g. price cannot be added to volume or time and so on.

Despite following the strategies mentioned above, it was found in our case studies that the number of features generated was still too large to carry out the necessary computations in a reasonable time. Furthermore, it was understood that the computational issues that arise due to memory limitations should be addressed in the feature generation framework implementation because the feature matrices were too large to be processed as single large matrices. Each grammar family produces a different number of feature expressions which are brought in Table 4.9 totaling 24,284. With 1,900 days under consideration, the feature matrix size was 1900×24284 for financial time-series (see Chap. 6).

4.5.2 Implementation Issues

This section discusses the implementation issues and secondary pruning strategies involved therein. Figure 4.4 depicts the feature generation flow which was used to

Table 4.9 Number of feature expressions generated by each family

Grammar family no	No of feature expressions
1	5,852
2	10,440
3	1,675
4	480
5	5,516
6	21
7	300
Total	24,284

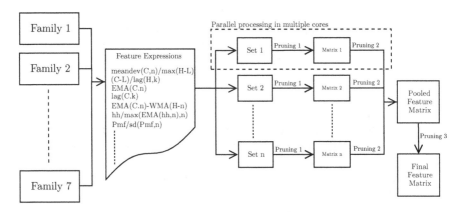

Fig. 4.4 Feature generation flow

generate features. Because of the design of grammar families, there were some feature expressions that were duplicated, i.e. some feature expressions were generated by multiple grammar families. These feature expressions were removed from the final feature expression list by simply searching for repeated formulae.

Each feature expression in the list was evaluated by substituting numerical values for the symbols of the expression. All possible n, k combinations of a particular feature expression were considered. The feature expression list was divided into 20 sets of roughly 1,000 features in each set. The feature expressions in each set were evaluated in a parallel environment using multiple cores since features can be independently evaluated.

For feature expressions having more than 10 permutations, "Pruning 1" (see Fig. 4.4) only selects the 10 best numerical features based on the information gain. If the number of permutations of a particular feature expression was less than 10, all permutations were selected.

It was also observed that some features had very low variance and not helpful in a regression. The feature matrices after the first prune were scaled and the features with low variance were removed by "Pruning 2".

All feature matrices were combined to form a large pool of features and the final feature matrix was achieved by further pruning. This final pruning, "Pruning 3", was carried out to remove features that were reciprocals of each other e.g. the feature generation framework treats EMA(H, 5) and 1/EMA(H, 5) as 2 distinct features but it is unclear if these 2 features add any additional information to the learning task. The MD5 hash for these 2 features are the same and this was exploited to keep only one of the features in the final matrix.

Continuing with the methodology of feature generation, ways of automatically generating only "good" feature subsets were also investigated. GE combines feature generation with feature selection by using the fitness function of the underlying GA to assess feature subset goodness using a wrapper approach. It is also shown that more degrees of freedom can be allowed in grammar by allowing recursive rules (see next section). This means that GE does not need any explicit feature pruning. Feature generation using GE is entailed in the next section.

4.6 Feature Generation Using GE

In this section, GE is proposed as a novel feature subset selection process. The grammar families in Table 4.10 were designed to generate *trend*, *momentum* and *volatility* type indicators for financial/electricity load time-series prediction in Chap. 6 and an initial population is evolved using GE to generate feature subsets that are better suited for prediction task at hand. Note the recursive rule (1.a) in the momentum family.

The theoretical background of GE was explained in Sect. 3.1. Figure 4.5 illustrates how the momentum grammar family maps a binary chromosome to a feature, namely the MACD technical indicator. The binary chromosome has 8 codons and each codon is converted to an integer resulting in the integer chromosome. The GE technique can also be directly used with integer chromosomes but using binary chromosomes leads to better genetic diversity. The start symbol is $\mathscr{S} = <expr>$. The first codon value is 222. Since there are 2 production rules (1.a), (1.b) that the non-terminal element $<expr>$ can take, (codon integer value) MOD (number of rules for the current non-terminal) is evaluated as $222\%2 = 0$ which points to rule (1.a). The derivation sequence of the figure can be understood in this manner noting that the left-most non-terminal is always expanded.

It was expected that a good mix of trend, momentum and volatility type formulae could result in better predictions. Therefore, gene partitions of individual chromosomes were mapped into different grammar families. Figure 4.6 depicts such an exemplary mapping of 5 features to 3 grammar families. The binary form of a chromosome is represented as a set of stacked genes for clarity. Each n bit long individual was dissected to N gene partitions where each gene was considered a feature and GE was independently applied on each partition, e.g. for an individual of 1,280 bits with 10 features $n = 1, 280, N = 10$.

It was already mentioned that GE uses standard genetic operators. Individuals were selected for recombination using the fitness proportionate selection (also known as

Table 4.10 High, low and close value based multi-family grammar

Moving average grammar family		
$\mathcal{N} = \{expr, der\text{-}var, base\text{-}var, pre\text{-}op, base\text{-}op, var\}$		
$\mathcal{T} = \{\texttt{delt}, \texttt{diff}, \texttt{ema}, \texttt{sma}, \texttt{wma}, \texttt{max}, \texttt{min}, \texttt{H}, \texttt{L}, \texttt{C}, \texttt{n}, \texttt{(}, \texttt{)}\}$		
$\mathcal{S} = \langle expr \rangle$		
Production rules: $\mathcal{R}$		
$\langle expr \rangle$	$::= \langle der\text{-}var \rangle$	(1.a)
	$\mid \langle base\text{-}var \rangle$	(1.b)
$\langle der\text{-}var \rangle$	$::= \langle pre\text{-}op \rangle(\langle base\text{-}var \rangle, \texttt{n})$	(2.a)
$\langle base\text{-}var \rangle$	$::= \langle base\text{-}op \rangle(\langle var \rangle)$	(3.a)
	$\mid \langle pre\text{-}op \rangle(\langle var \rangle, \texttt{n})$	(3.b)
	$\mid \langle var \rangle$	(3.c)
$\langle pre\text{-}op \rangle$	$::= \texttt{ema} \mid \texttt{sma} \mid \texttt{wma} \mid \texttt{max} \mid \texttt{min}$	(4.a), (4.b), (4.c), (4.d), (4.e)
$\langle base\text{-}op \rangle$	$::= \texttt{delt} \mid \texttt{diff}$	(5.a), (5.b)
$\langle var \rangle$	$::= \texttt{H} \mid \texttt{L} \mid \texttt{C}$	(6.a), (6.b), (6.c)
Momentum grammar family		
$\mathcal{N} = \{expr, var\text{-}op, op, var\}$		
$\mathcal{T} = \{\div, -, \texttt{delt}, \texttt{lag}, \texttt{ema}, \texttt{H}, \texttt{L}, \texttt{C}, \texttt{n}, \texttt{k}, \texttt{(}, \texttt{)}\}$		
$\mathcal{S} = \langle expr \rangle$		
Production rules: $\mathcal{R}$		
$\langle expr \rangle$	$::= (\langle expr \rangle)\langle op \rangle(\langle expr \rangle) \mid \langle var\text{-}op \rangle$	(1.a), (1.b)
$\langle var\text{-}op \rangle$	$::= \texttt{lag}(\langle var \rangle, \texttt{k}) \mid \texttt{ema}(\langle var \rangle, \texttt{n})$	(2.a), (2.b)
$\langle op \rangle$	$::= \div \mid -$	(3.a), (3.b)
$\langle var \rangle$	$::= \texttt{H} \mid \texttt{L} \mid \texttt{C} \mid \texttt{delt}(\texttt{H})$	(4.a), (4.b), (4.c), (4.d)
Volatility grammar family		
$\mathcal{N} = \{expr, var\text{-}op, op, var\}$		
$\mathcal{T} = \{+, -, \texttt{abs}, \texttt{ema}, \texttt{sd}, \texttt{meandev}, \texttt{H}, \texttt{L}, \texttt{C}, \texttt{n}, \texttt{(}, \texttt{)}\}$		
$\mathcal{S} = \langle expr \rangle$		
Production rules: $\mathcal{R}$		
$\langle expr \rangle$	$::= \texttt{ema}(\langle var\text{-}op \rangle, \texttt{n}) + \texttt{sd}(\langle var\text{-}op \rangle)$	(1.a)
	$\mid \texttt{ema}(\langle var\text{-}op \rangle, \texttt{n}) - \texttt{sd}(\langle var\text{-}op \rangle)$	(1.b)
$\langle var\text{-}op \rangle$	$::= \texttt{abs}(\langle var \rangle) \mid \texttt{meandev}(\langle var \rangle) \mid \langle var \rangle$	(2.a), (2.b), (2.c)
$\langle var \rangle$	$::= \texttt{H-L} \mid \texttt{H-C} \mid \texttt{C-L}$	(3.a), (3.b), (3.c)
	$\mid \texttt{H} \mid \texttt{L} \mid \texttt{C}$	(3.d), (3.e), (3.f)

roulette wheel selection). Random mutation was employed and a modified version of multiple-point crossover in Fig. 4.7 was used. The stacked gene was partitioned based on randomly selected cross-over points and the sections were interchanged to produce the children. This crossover ensured that inter-grammar-family crossover is avoided. Elite individuals were passed to the next generation without mutation or crossover.

A unique advantage is the ability to exploit the GE fitness function as a mechanism to penalize bad feature subsets and evolve the population towards "good" ones,

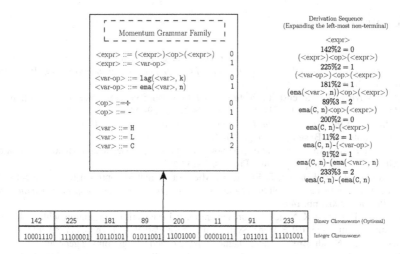

Fig. 4.5 GE mapping from a linear binary (or integer) chromosome

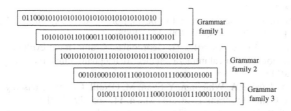

Fig. 4.6 Gene partition mapping example for 5 features

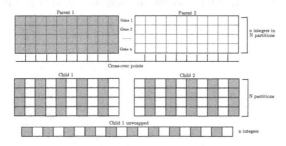

Fig. 4.7 Proposed multiple-point crossover

leading to better predictions. Different parameters can be controlled to get feature subsets with desired properties (feature complexity, number of parameters in the features, feature expression character length etc.). This crafting of GE fitness function as a wrapper is described in more detail in Sect. 5.3.2.

References

1. D. Eads, K. Glocer, S. Perkins, J. Theiler, Grammar-guided feature extraction for time series classification, in *Proceedings of the 9th Annual Conference on Neural Information Processing Systems*, 2005
2. S. Markovitch, D. Rosenstein, Feature generation using general constructor functions, in *Machine Learning*, vol. 49 (The MIT Press, 2002), pp. 59–98
3. M. Sipser, in *Introduction to the Theory of Computation, Chapter 2*. Context-Free Grammars (PWS Publishing, Boston, 1997), pp. 91–122
4. F. Pachet, P. Roy, Analytical features: a knowledge-based approach to audio feature generation. EURASIP J. Audio Speech Music Process. 1–23 (2009)
5. O. Ritthof, R. Klinkenberg, S. Fischer, I. Mierswa, A hybrid approach to feature selection and generation using an evolutionary algorithm, in *2002 U.K. Workshop on Computational Intelligence*, 2002, pp. 147–154
6. H. Guo, L.B. Jack, A.K. Nandi, Feature generation using genetic programming with application to fault classification. IEEE Trans. Syst. Man Cybern. Part B: Cybern. **35**(1), 89–99 (2005)
7. H. Guo, A.K. Nandi, Breast cancer diagnosis using genetic programming generated feature, in *2005 IEEE Workshop on Machine Learning for Signal Processing*, 2005, pp. 215–220
8. R. Islamaj, L. Getoor, W. John Wilbur, A feature generation algorithm for sequences with application to splice-site prediction, in *Knowledge Discovery in Databases: PKDD 2006* (Springer, 2006), pp. 553–560
9. K. Krawiec, B. Bhanu, Visual learning by coevolutionary feature synthesis. IEEE Trans. Syst. Man Cybern. Part B: Cybern. **35**(3), 409–425 (2005)
10. O. Smart, I.G. Tsoulos, D. Gavrilis, G. Georgoulas, Grammatical evolution for features of epileptic oscillations in clinical intracranial electroencephalograms. Expert Syst. Appl. **38**(8), 9991–9999 (2011)
11. R. McKay, N.X. Hoai, P.A. Whigham, Y. Shan, M. ONeill, Grammar-based genetic programming: a survey. Genet. Program. Evolvable Mach. **11**, 365–396 (2010)
12. S.-C. Huang, T.-K. Wu, Integrating GA-based time-scale feature extractions with SVMs for stock index forecasting. Expert Syst. Appl. **35**(4), 2080–2088 (2008)
13. R.W. Schabacker, *Stock Market Theory and Practice* (BC Forbes, New York, 1930)
14. R.D. Edwards, J. Magee, W.H.C. Bassetti, *Technical Analysis of Stock Trends* (CRC, Boca Raton, 1948)

Chapter 5
Application of Grammar Framework to Time-Series Prediction

Abstract The previous chapter presented an approach to generate a large number of features using an expert-defined grammar framework. This chapter proceeds to investigate ways to explore such large feature spaces to extract the best features for prediction, i.e. feature selection (FS). Since the proposed framework involves the generation of a large pool of features, there can be redundant and irrelevant features. Therefore, FS is as equally important as feature generation. Several FS and feature extraction techniques can be explored to determine the best approach to discover "good" feature subsets for particular ML algorithms in different applications. A hybrid feature selection and generation algorithm using grammatical evolution is described as a technique to avoid selective feature pruning by crafting the fitness function to penalise *bad* feature subsets. The chapter also describes how ML algorithms were used to predict time-series using the sliding window technique, data partitioning, model selection and parameter tuning.

Keywords Financial time-series prediction · Electricity load time-series prediction · Data preprocessing · Model selection · Parameter tuning

5.1 Expert-Suggested Features

When applying ML techniques to a time-series prediction task, the input features are usually chosen by domain experts. Different application domains adopt features of different styles, e.g. financial time-series prediction applications use technical indicators, electricity load time-series prediction applications use a history window, temperature, calendar information, etc. The common input features for financial time-series and electricity load time-series prediction are presented in Tables 5.1 and 5.2, respectively. These common features were used as a basis to construct grammar structures for each application in Chap. 6. By doing so, it was possible to generate commonly used features and many other potentially "good" feature subsets in a large candidate feature pool. Similar input features are used across studies with different learners with seldom changes in input feature formulation choice. For example, in the works of Table 5.1, the authors identify a subset of the standard technical indicators

© The Author(s) 2015   51
A.M. De Silva and P.H.W. Leong, *Grammar-Based Feature Generation*
for Time-Series Prediction, SpringerBriefs in Computational Intelligence,
DOI 10.1007/978-981-287-411-5_5

Table 5.1 Common technical indicators used as input features in financial time-series prediction

Technical indicators used as features	# of TIs	Data	Method	References
K, D, Slow D, MOM, ROC, R, ADO, DIS, OSCP, RSI, CCI	13	KOSPI	SVM	[1]
O, H, L, C, V, RSI, WMA	7	N225, TAIEX	ICA + SVM	[2]
SMA, RSI, PSY, MOM, D, VR, OBV, DIS, ROC	9	KOSPI	GA	[3]
C, K, D, Slow D, ROC, MOM, SMA, σ, σ ratio, EMA, MACD, ADO, DIS, OSCP, CCI, RSI	18	S&P 500	SVM + GA	[4]
SMA, BIAS, RSI, K, D, MACD, PSY, V	8	TSEC stocks	K-means + FDT + GA	[5]
SMA, EMA, Projection oscillator, MACD, Qstick, TRIX, etc	54	DJI, S&P 500	ANN	[6]
SMA, RSI, K, D, MACD, R, PSY, $\Gamma^\pm$, BIAS, VR, A ratio, B ratio	13	FITX	SOM + SVM	[7]
Returns, differences of returns, oscillators, SMA, EMA	42	BEL 20	KPCA + RBF-NN	[8]
EMA, ADO, K, RSI, ROC, Price accelerations	10	Stocks	ABFO + BFO	[9]
ΔC, BB, RSI, K, M, CLV, MFI, R	10	Stocks	KPCA + SVM	[10]
OBV, SMA, BIAS, PSY, Average stock yield, C	12	Stocks	AFSA + NN	[11]
SMA, C	6	Forex	ARIMA + ANN	[12]

OBV On-balance volume, *VR* Volume ratio, TRIX is the percentage change of the triple-smoothed moving average of the closing price, Qstick is another oscillator and *PSY* Psychological stability of investors. $\Gamma^\pm$ are the directional movement indicators

KOSPI Korea Composite Stock Price Index, *N*225 Japanese Nikkei index, *TAIEX* Taiwan Stock Exchange Capitalization Weighted Stock Index, *S&P* 500 US standard & poor's index, *TSEC* Taiwan Stock Exchange Corporation, *DJI* Dow Jones index, *FITX* Taiwan index futures, *BEL* 20 Belgian stock index

ICA Independent component analysis, *GA* Genetic algorithm, *FDT* Fuzzy decision tree, *ANN* Artificial neural network, *SOM* Self-organising map, *KPCA* Kernel principal component analysis, *RBF-NN* Radial basis function neural network, *ASFA* Adaptive fish swarm algorithm, *ABFO* Adaptive bacterial foraging optimization and *ARIMA* Autoregressive integrated moving average

that are deemed informative. In each case, the number of technical indicators selected is not the same with some overlap between different works. However, the choice is generally ad hoc.

5.1.1 Stock Index Time-Series Prediction

The analysis of financial time-series is of critical importance to a wide range of business such as corporate banks, broker firms, individual investors and other organisations. Such analysis is used for the hedging of risk, speculative and algorithmic trading, portfolio management, planning and other activities. The accuracy of financial predictions and the speed at which the predictions can be obtained is of great interest. The efficiency and complexity of financial markets, however, make reliable learning of financial time-series an extremely challenging problem.

Seminal work by Fama [19] proposes the random walk hypothesis on stock market patterns which states "The main conclusion will be that the data seem to present consistent and strong support for the model. This implies, of course, that chart reading, though perhaps an interesting pastime, is of no real value to the stock market investor". Many subsequent studies agree with this theory, while almost an equal number disagree. Lo et al. [20] used nonparametric kernel regression and concluded that "several technical indicators do provide incremental information and may have some practical value". This has always been a controversial topic, and it is generally believed that some predictability can be achieved over some time periods. Based on this belief, many studies are still conducted on pure financial time-series prediction.

The purpose of our research was to investigate the success achieved by using effective, but seemingly unconventional feature combinations which are not apparent to human experts. Hence, our case study in Sect. 6.1 does not attempt to make advances in the state-of-the-art financial time-series prediction systems. Nevertheless, it is straightforward for any state-of-the-art financial prediction system to make use of the proposed approach as a preprocessing stage to select feature subsets.

A standard approach in the financial time-series literature is to choose a set of standard technical indicators and/or external economic factors as input features. Table 5.1 presents the symbols of technical indicators used, number of technical indicators used and method in 12 studies. References [1, 2, 7] used standard technical indicators as features in the support vector machine (SVM) to predict financial time-series. In a similar fashion, standard technical indicators have been used as features for neural networks [1, 6, 12]. References [8, 10] used kernel principal component analysis (KPCA) to reduce the dimensionality of the feature space formed by standard technical indicators and used SVM and neural network to predict stock prices. Other techniques such as SVMs optimised using genetic algorithm (GA) [3], neural networks optimised using adaptive fish swarm algorithm (AFSA) [11], boosted experts using GA [3], adaptive bacterial foraging optimization (ABFO) [9] have also been used to model financial time-series. All these applications manually selected a set of standard technical indicators as features for different ML techniques.

In the above-mentioned works, the authors identify a subset of the standard technical indicators that are deemed informative. In each case, the technical indicator, number of technical indicators selected and importantly, the technical indicator parameters were not the same, the choice generally being ad hoc. It would therefore be useful to have a framework that can automatically generate interesting feature combinations by systematically guiding the generation of a pool of candidate features that have a meaningful interpretation.

5.1.2 Electricity Load Demand Time-Series Prediction

Accurate load prediction plays a major role in distribution system investments and electricity load planning and management strategies. Bunn and Farmer [21] pointed that a 1 % increase in prediction error implied a £10 million increase in operating costs. While overestimation results in an unnecessary spinning reserve and undesired excess supply, underestimation causes failure in providing sufficient reserve and implies high costs per peaking unit. Therefore, it is desirable to predict electricity load demand accurately. It is shown that using decomposed signals is more effective in ML-based electricity load prediction than using raw time-series signals [22]. Fourier series can be used to decompose electricity load time-series, but it is far useful when the time-series is stationary. Empirical mode decomposition (EMD) and wavelet transforms are the most popular approaches. In predicting nonstationary time-series, such multi-resolution decomposition techniques can be used for elucidating complex relationships [23, 24]. The wavelet transform can produce a good local representation of a signal in both time and frequency domains and is not restrained by the assumption of stationarity. EMD is also an alternative to wavelet transform but EMD-based feature generation. History windows and first differences of the raw time-series and decomposed components seem to be popular expert hand-picked features in electricity load time-series prediction (see Table 5.2).

Table 5.2 Common input features used in short-term electricity load time-series prediction

Features	Method	References
Previous hour load history windows	SVM	[13, 14]
Previous week's hourly load information, temperature information, calendar information	SVM	[15]
Previous load information and differenced values, Daubechies wavelets (multi-resolution analysis of hourly load up to 168 h ago) and differenced values, temperature information	ANN	[16]
Wavelets (multi-resolution analysis of hourly load up to 500 h ago), temperature information	ANN, GA	[17]
Empirical mode decomposition (multi-resolution analysis of hourly load), temperature information	SVM	[18]

5.1.3 Better Feature Combinations Than Common Input Features?

From Tables 5.1 and 5.2, it can be seen that applications tend to use a similar set of features with ad hoc parameter values. As introduced in Chap. 1, we were intrigued by the question, "can we find feature combinations better than a human expert-selected features for a given ML architecture?". It can be argued that having a better observation language can lead to a better hypothesis language leading to a better solution hypothesis (see Chap. 1). By searching for better input feature combinations generated via an expert-defined grammar framework, we attempted to answer this question under certain conditions.

5.2 Data Partitioning for Time-Series Prediction

In a typical application of ML to time-series prediction, it is common practice to divide the time-series into training, validation and testing (out-of-sample) sets (see Fig. 5.1). The training set is used to construct a model. The model is supposed to fit the samples in the training set. The validation set is used to evaluate the generalisation ability of the trained model. The model parameters are tuned such that the model performs satisfactorily on the validation set. The final evaluation of the model is based on its performance on the testing set. No changes to the model should be made by repeatedly performing a series of train–validation–test steps and adjusting the model parameters and features based on the model performance on the testing set. This is a form of "peeking" (see Sect. 5.4).

The training set is the largest in size and the validation set is usually 10–30 % of the training set size. The testing set should have sufficient samples to evaluate the trained model (in our case studies, the same sample size is used for both the validation and the testing set). The testing set should consist of the most recent contiguous observations.

A more rigorous approach, called the sliding or moving window (also known as walk-forward testing), is a form of online training because the model is frequently retrained. The number of samples in the testing set determines the retraining frequency. For one-step ahead predictions, this means that the model can be retrained after every prediction. The data are divided into a series of overlapping training–validation–testing sets. The typical training–validation–testing concept is still present, but now only the most recent observations are used to construct models.

Fig. 5.1 Training, validation and testing samples in a typical ML application

Fig. 5.2 The sliding
window approach for daily
model retraining

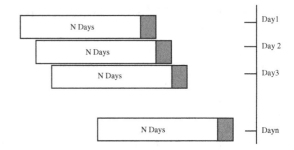

This is depicted in Fig. 5.2 for daily model retraining. Frequent retraining is more
time-consuming but allows the models to adapt more quickly to the nonstationarity
in time-series.

5.2.1 Data Preprocessing

The feature matrix is "cleaned" before being used by any ML model. The missing values are interpolated, the infinite values are replaced with a finite maximum
threshold, and the outliers are usually removed.

The training and validation sets are scaled together since the purpose of the testing
set is to determine the ability of the learner to generalise. However, the testing set
should not be scaled with either the training or validation sets since this biases the
integrity of the validation set as a final and independent check on the model.

The self-organising map (SOM) makes use of a set of prototype vectors represent-
ing the dataset and performs a topology preserving projection of the prototypes from
input space to a low-dimensional grid [25]. This grid can be used as a visualisation
surface to identify clusters in the original data in an unsupervised manner. This has
been exploited in many time-series prediction applications such as electricity load
time-series prediction to identify similar data patterns, e.g. if the objective is to predict the electricity load in January, a SOM can be used to identify periods (months)
that show similar load patterns to January. Training a model only on similar data
patterns can not only lead to better prediction but also can reduce the model training
time.

5.2.2 Model Selection and Parameter Tuning

Model selection and parameter search is critical to the performance of any ML
algorithm. For a support vector machine (SVM), the kernel functionis such a

parameter that should be chosen to maximise the performance. Radial basis functions (RBF) which nonlinearly map the samples to the high-dimensional space is generally known to work well in many applications. The RBF with the kernel parameter γ takes the form $K(x_i, x_j) = exp(-\gamma\|x_i - x_j\|^2)$, where x_i, x_j are the input training vectors for $i, j = 1, 2, \ldots, m$. For a SVM using a RBF kernel, the parameters C, γ needs to be tuned where C is the penalty term. Improper parameter selection can lead to over/under fitting [26]. Cross-validation via parallel grid search, genetic algorithms, random search, heuristics search and inference of model parameters within the Bayesian evidence framework are some parameter search techniques. The performance of different parameter combinations is assessed by the learner performance, e.g. mean-squared error (MSE).

The case studies in this brief use parameter evaluation via a parallel grid search on the validation data. Parameter tuning using the validation data prevents the overfitting problem. The final performance of the learner is evaluated using the best parameters in the validation phase.

K-fold cross-validation, two-fold cross-validation, leave-one-out cross-validation and repeated random sampling cross-validation are some popular cross-validation techniques. Time-series cross-validation is slightly different because the data are not independent, and leaving an observation out does not remove all the associated information due to the correlations with other observations and can be done as suggested by Hyndman [27],

1. Fit the model to the data $y_1, \ldots, y_t$ and let $\hat{y}_{t+1}$ denote the prediction of the next observation. Then, compute the error RMSE as $e_t^* = \sum_{i=1}^{n}(y_i - \hat{y}_i)^2$
2. Repeat step 1 for $t = m, \ldots, n - 1$ where m is the minimum number of observation needed for fitting the model
3. Compute the average RMSE from $e_{m+1}^*, \ldots, e_n^*$

5.3   Feature Selection Techniques

As mentioned in Chap. 4, the CFG-based framework generates a large number of features parametrised by the look-back period N, window size n and lag k in time-series prediction. Some features have no parameters while others have one or both. It is important to select the appropriate (N, n, k) for each feature to obtain the most informative features, e.g. some specific lags (k) of a feature can have much important information than other lags of the same feature, a feature with a shorter window (n) can have better predictive power than the same feature with a longer window.

A range of FS techniques were explored in our case studies in Chap. 6 to compare the performance. The filter FS techniques (see Sect. 2.2.1) used were information gain, maximum relevance minimum redundancy (mRMR), correlation and Relief. The individual feature goodness for each feature was assessed against the target variable, e.g. the closing price of stock index time-series. Once the features were ranked, an appropriate number of features was used for the prediction task. The advantage of filters is that they are extremely fast compared to wrapper-based FS. As already mentioned, filter-ranked features can have redundancies since similar features have equal rankings (weights). Furthermore, individual feature goodness does not translate to better predictive power when used in ML algorithms. Therefore, it is hard to determine the optimal number of features to use. PCA and wrapper approaches lead to better results. The use of grammatical evolution (GE) as a wrapper-based hybrid feature generation and selection technique is explained in Sect. 5.3.2 and showed promise.

5.3.1 Dimensionality Reduction Using PCA

Principal component analysis (PCA) has been proposed as an effective FS technique in many applications. The generated features should be first ranked according to different filter criteria and the number of features to perform PCA should be chosen based on a sharp cut-off threshold. PCA can be applied on this chosen number of features. Different numbers of principal components accounting for different threshold in the variance of original data can be used as the features to compare performance.

   PCA for large matrices is a time-consuming process. When using the sliding window technique for PCA predictions, m models should be constructed, where m is the testing set size. If the size of the parameter grid is 100×100, the total number of models becomes $100 \times 100 \times m$ and an equal number of PCAs should be performed. Since the calculations are required to be performed within a reasonable time, the PCA components (transformed feature matrix, rotational matrix and scaling matrix) on each training window can be stored and reused as illustrated in Fig. 5.3. The testing features can be scaled and rotated using the rotational and scaling matrices of the training data as shown. This approach requires only m PCAs to be performed hence less time-consuming. A review of PCA theory is omitted in this brief.

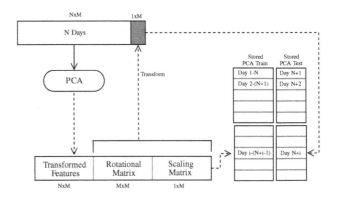

Fig. 5.3 PCA transformations for the ith day

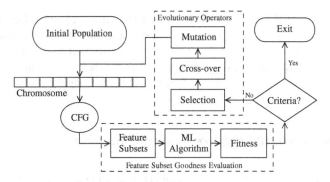

Fig. 5.4   System architecture for integer GA-based FS

5.3.2 *Feature Selection Using Integer Genetic Algorithms*

The generic details of canonical GAs were presented in Sect. 3.2. This section elaborates specificities of adopting integer GA for FS.

Figure 5.4 shows a possible architecture for integer GA-based FS used in our work. Chromosomes can be selected from the population which are then mapped to feature subsets using the defined CFG. These subsets can be evaluated and the best subsets can be selected by a wrapper approach using a particular learner architecture with fixed parameter settings. The population can be evolved such that the learner architecture has a minimum cross-validation error.

The initially generated feature space is often too large for the search algorithms to converge in a reasonable time. Therefore, when GE is not used as a hybrid feature generation and selection technique, it is suggested the feature space to be first shrunken by using mRMR criterion to rank and select top features. This shrunken feature space can be used to seek better performing feature subsets using the wrapper approach.

In order to quickly discover better feature subsets, specific chromosomes can be placed in the initial population which are known to work well in general, e.g. standard technical indicators in financial time-series prediction. This is possible since the rule sequence to generate a specific technical indicator (or a feature in general) is known. This ensures that the initial population is healthy and encourages the generation of high performing feature subsets. The rest of the population can consist of randomly generated individuals.

Consider the integer chromosome $[302|237|2|451|\dots|871|23|657|77|549]$. Each gene represents a feature number. For example, if the feature space size is 1,000 features, the chromosome says to select the features 302, 237, 2 and so on. Integer GA mutation can be done according to the criterion $x(1 + \mathtt{rand}(-0.05, 0.05))$, where x is the current codon and $\mathtt{rand}$ is a function which generates a random number between -0.05 and 0.05 (as opposed to random bit flipping in canonical GA). Crossover and selection criteria can be canonical. The feature subset size can be varied by varying the chromosome length, i.e. the number of genes. Because of the inherent random nature of the GA algorithm, it is usually run multiple times before drawing any conclusions. At least 10 trials were performed in the work involved with this brief, and a final result was reported as the best, worst and the average performance for the 10 runs as in standard practice (see Chap. 6).

5.3.3 Wrapper-Based Feature Subset Evaluation

This section explains wrapper-based FS and in particular how the GE fitness function can be crafted to select good feature subsets. The feature subset evaluation steps used in electricity load time-series prediction using GE in our case study of electricity load time-series prediction (see Sect. 6.2) are shown in Algorithm 4. The individuals with the lowest score were deemed to be the best. $e(Y_k)$ is the MAPE (mean absolute percentage error), given by $\frac{100}{n} \sum_{i=1}^{n} \left| \frac{y_i - \hat{y}_i}{y_i} \right|$ where n is the number of predictions, y is the target, and $\hat{y}$ is the predicted value. To choose robust features providing good generalisation, time-series cross-validation was used (see Sect. 5.2.2). The lines 6–8 correspond to time-series cross-validation, and N was chosen to be 5, i.e. fivefold time-series cross-validation. $q(\cdot)$ is an assessment of the symbolic feature expression complexity, taking into account the length, the number of operations and preoperations. Since GE-generated feature subsets can include nonterminal expressions, $c(\cdot)$ was calculated based on the number of nonterminal elements N_{NT} in subset Y_k (a subset containing many nonterminal elements was considered as poor). When GE was not used, the genetic feature subset evaluation steps are specified in Algorithm 5.

Algorithm 4 Feature subset evaluation procedure using GE

Input current integer chromosome
 1: Extract individual genes from the current chromosome
 2: Map genes into grammar families (if necessary)
 3: Derive symbolic features
 4: Generate numeric feature subset Y_k for each time-stamp
 5: Append calendar information to Y_k (see Chap. 6)
 6: **for** i in 1:N **do**
 7:     err[i] = ML algorithm MAPE in validation sample i
 8: **end for**
 9: Calculate the average MAPE $e(Y_k) = \text{mean(err)}$
10: Calculate the final score $J(Y_k) = e(Y_k) + q(Y_k) + c(N_{NT})$
Output $J(Y_k)$

Algorithm 5 Feature subset evaluation procedure

Input current symbolic features
 1: Generate numeric feature subset Y_k for each time-stamp
 2: Append selected expert-suggested features to Y_k
 3: **for** i in 1:N **do**
 4:    err$[i]$ = ML algorithm performance metric in validation sample i
 5: **end for**
 6: Calculate the average performance metric $e(Y_k) =$ mean(err)
Output $e(Y_k)$

5.4 Avoiding Peeking

We use the term "peeking" to scenarios where future data are *mistakenly* used to get an insight into the future behaviour. This section explains 2 techniques that can be implemented to ensure that this mistake is avoided.

(i) It must be ensured that in prediction or parameter tuning, the implementation function had no access to future data. The safest way to avoid peeking when predicting x_{t+1} is to ensure that all training samples are time-stamped and the prediction/tuning function to return an error/terminate if it encounters any samples with time-stamps after x_t. A wavelet decomposition should also be carried out in an incremental manner to avoid peeking as described in Appendix A.

(ii) Even if the time-stamp check is performed, there is still the risk of peeking involved in feature construction. It should be ensured that a particular feature is constructed with only past values. A validity test was devised to confirm this. This test compared features constructed from all data up to time t with features constructed from all data up to time $t + k$ for some interval k. The numeric evaluation of the constructed features in two cases should be identical for the period t. If this is not the case, it implies that in constructing features for the period t, the data in interval k have been used in some way.

References

1. K.-J. Kim, Financial time series forecasting using support vector machines. Neurocomputing **55**(1), 307–319 (2003)
2. C.-J. Lu, T.-S. Lee, C.-C. Chiu, Financial time series forecasting using independent component analysis and support vector regression. Decis. Support Syst. **47**(2), 115–125 (2009)
3. M.-J. Kim, S.-H. Min, I. Han, An evolutionary approach to the combination of multiple classifiers to predict a stock price index. Expert Syst. Appl. **31**(2), 241–247 (2006)
4. L. Yu, S. Wang, K.K. Lai, Mining stock market tendency using ga-based support vector machines. Lect. Notes Comput. Sci. **3828**, 336–345 (2005)
5. R.K. Lai, C.-Y. Fan, W.H. Huang, P.-C. Chang, Evolving and clustering fuzzy decision tree for financial time series data forecasting. Expert Syst. Appl. **36**(2), 3761–3773 (2009)

6. A. Zapranis, Testing the random walk hypothesis with neural networks, *Artificial Neural Networks ICANN 2006*, Lect. Notes in Comput. Sci. vol 4132 (Springer, Berlin Heidelberg, 2006), pp. 664–671

7. C.-L. Huang, C.-Y. Tsai, A hybrid SOFM-SVR with a filter-based feature selection for stock market forecasting. Expert Syst. Appl. **36**(2), 1529–1539 (2009)

8. A. Lendasse, E. de Bodt, V. Wertz, M. Verleysen, Non-linear financial time series forecasting— application to the BEL-20 stock market index. Eur. J. Econ. Soc. Syst. **14**(1), 81–91 (2000)

9. R. Majhi, G. Panda, B. Majhi, G. Sahoo, Efficient prediction of stock market indices using adaptive bacterial foraging optimization and BFO based techniques. Expert Syst. Appl. **36**(6), 10097–10104 (2009)

10. H. Ince, T.B. Trafalis, *Kernel principal component analysis and support vector machines for stock price prediction*, in IEEE International Joint Conference Neural Networks, vol. 3, 2053–2058 (2004)

11. W. Shen, X. Guo, C. Wu, D. Wu, Forecasting stock indices using radial basis function neural networks optimized by artificial fish swarm algorithm. Knowl. Based Syst. **24**(3), 378–385 (2011)

12. J. Kamruzzaman, R.A. Sarker, Forecasting of currency exchange rates using ANN: a case study, in *Proceedings of the 2003 International Conference on Neural Networks and Signal Processing*, vol. 1, pp. 793–797 (2003)

13. B.-J. Chen, M.-W. Chang et al., Load forecasting using support vector machines: a study on EUNITE competition 2001. IEEE Trans. Power Syst. **19**(4), 1821–1830 (2004)

14. M. Mohandes, Support vector machines for short-term electrical load forecasting. Int. J. Energy Res. **26**(4), 335–345 (2002)

15. M. Espinoza, J. Suykens, B. De Moor, Load forecasting using fixed-size least squares support vector machines. Comput. Int. Bioinspired Syst. **1**, 488–527 (2005)

16. A.J.R. Reis, A.P.A. da Silva, Feature extraction via multiresolution analysis for short-term load forecasting. IEEE Trans. Power Syst. **20**(1), 189–198 (2005)

17. J. Yao, C.L. Tan, A case study on using neural networks to perform technical forecasting of forex. Neurocomputing **34**(14), 79–98 (2000)

18. L. Ghelardoni, A. Ghio, D. Anguita, Energy load forecasting using empirical mode decomposition and support vector regression. Smart Grid IEEE Trans. **4**(1), 549–556 (2013)

19. E.F. Fama, The behavior of stock-market prices. J. Bus. **38**(1), 34–105 (1965)

20. A.W. Lo, H. Mamaysky, J. Wang, Foundations of technical analysis: computational algorithms, statistical inference, and empirical implementation. J. Finance **55**(4), 1705–1770 (2000)

21. D. Bunn, E. Dillon Farmer, Comparative models for electrical load forecasting (Wiley and Sons Inc., New York, 1985)

22. C.-J. Yu, Y.-Y. He, T.-F. Quan, Frequency spectrum prediction method based on EMD and SVR, in *Eighth International Conference on Intelligent Systems Design and Applications, ISDA '08*, vol. 3, pp. 39–44 (2008)

23. D. Benaouda, F. Murtagh, J.-L. Starck, O. Renaud, Wavelet-based nonlinear multiscale decomposition model for electricity load forecasting. Neurocomputing **70**(1), 139–154 (2006)

24. Z.R. Struzik, Wavelet methods in (financial) time-series processing. Phys. A: Stat. Mech. Appl. **296**(1), 307–319 (2001)

25. T. Kohonen, *Self-organizing maps*, vol 30 (Springer, Berlin, 2001)

26. F.E.H. Tay, L. Cao, Application of support vector machines in financial time series forecasting. Omega **29**(4), 309–317 (2001)

27. R.J. Hyndman, Why every statistician should know about cross-validation, (2010). http://robjhyndman.com/hyndsight/crossvalidation/

Chapter 6
Case Studies

Abstract This chapter presents how the performance of selected learning algorithms is affected by using various input feature combinations with different parameter values. The results are discussed under three sections for stock market index time-series, electricity load demand time-series and foreign exchange client trade volume time-series. The effectiveness of the feature subsets selected using the proposed approach was compared to the performance of the same algorithms trained using commonly (and widely) used input features and other benchmarks. By "good" features, a reference is made to features that are "good for a particular ML algorithm architecture/configuration" because it is difficult to define universally good features.

Keywords Stock index time-series prediction · Electricity load time-series prediction · Foreign-exchange client trade volume · Risk hedging

6.1 Predicting Stock Indices

A stock market index reflects the movement average of many individual stocks rather than the movement of a single stock. It can be believed that stock market indices are easier to model and predict than individual stocks, which can be very volatile.

Numerous studies, some which are summarised in Table 5.1, use technical indicators as input features for prediction. This relies on the past events of the time-series captured using technical indicators repeating to produce reliable predictions and is known as *technical analysis*. This section explores the effectiveness of using such standard technical indicators as features and attempts to discover numeric feature combinations with different parameter values that can give better predictions.

The system performance was assessed on the daily closing prices of leading financial stock indices in Table 6.1. The data were downloaded from "Yahoo! Finance" using the `quantmod` package in R [1] and comprise of daily recordings between 16 October 1998 and 19 June 2006 (1,900 trading days). The data preprocessing and parameter tuning were carried out as described in Chap. 5.

© The Author(s) 2015
A.M. De Silva and P.H.W. Leong, *Grammar-Based Feature Generation for Time-Series Prediction*, SpringerBriefs in Computational Intelligence,
DOI 10.1007/978-981-287-411-5_6

Table 6.1 Major world financial indices

Symbol	Index name	Listed exchange
AORD	All ordinaries index	Australia Stock Exchange
FTSE	FTSE-100 Index	London Stock Exchange
GDAXI	DAX index	Germany Stock Exchange
GSPC	S&P-500 (Standard and Poor's) Index	–
HSI	Hang Seng Index	Hong Kong Stock Exchange
TWII	Taiwan Weighted Stock Index	Taiwan Stock Exchange
NDX	NASDAQ-100 Index	NASDAQ stock market
N225	NIKKEI 225 Index	Osaka Securities Exchange
SSEC	Shanghai Stock Exchange Composite index	Shanghai Stock Exchange
SSMI	Swiss Market Index	Six Swiss Exchange

The predictability of indices was first assessed using widely used standard technical indicators in Table 4.5 with the exception of SMA and WMA because EMA is much popular. The grammar families in Tables 4.6 and 4.7, B.1–B.5 were then used to generate a large pool of features, and different feature selection (FS) criteria in Sect. 5.3 were used to select *custom technical indicator* combinations with different parameter values. For a fair comparison, the number of features used in both cases was the same, 25 features.

Support vector regression (SVR) and back-propagation neural network (BPNN) were used as ML methods and the experiment configuration details are in Table 6.2. SVM with a Gaussian kernel was found to work best, and the parameters were selected by a grid search parallelised on multiple cores using the snowfall package in R [2]

Table 6.2 Financial index prediction experiment configuration

Prediction horizon	One-day ahead
Training period	1998-10-16 to 2004-11-11 (1,500 days)
Validation period	2004-11-12 to 2005-09-01 (200 days)
Testing period	2005-09-02 to 2006-06-19 (200 days)
Support vector regression + Gaussian kernel	
SVM cost range (C)	$\{1, 5i, i = 1, 2, \ldots, 40\}$
SVM kernel hyper-parameter range (γ)	100 values spread out between $\{0.00001, 1\}$
SVM hyper-parameter (ε)	$0.1, 0.01, 0.001$
Back-propagation neural network	
Number of layers	3 (input, output and hidden)
Number of nodes in hidden layer	8
Other learning parameters	Optimised for best validation performance[a]

[a] The learning rate, momentum and other learning parameters were automatically decided by the MATLAB neural network toolbox to get the best performance by using the least number of epochs

Table 6.3 Steps in the financial time-series prediction experiments

(1) Select a ML algorithm, e.g. SVM

(2) Select well-known (and widely used) technical indicators (25 standard technical indicators as described above)

(3) Use the validation data to select an appropriate kernel (linear, polynomial and Gaussian kernels were considered) for the 25 standard technical indicators

(4) Optimise kernel parameters for the 25 standard technical indicators—Result 1

(5) Generate grammar features

(6) Apply pruning strategies to prune the feature space

(7) Use mRMR to shrink the grammar feature space

(8) Use integer genetic algorithm (GA) with fixed parameter-valued SVM as a wrapper to select feature subsets from the shrunken space

(9) Re-optimise the kernel for the integer GA-selected feature subset—Result 2

(see Appendix C Sect. C.2 for selected parameters). The SVM was implemented using the `e1071` package [3]. For the BPNN, a 3-layer architecture with 8 hidden nodes was implemented [4–9] using the neural network toolbox in MATLAB®.

The steps involved in the financial time-series prediction experiments are in Table 6.3.

To gauge the performance of the ML techniques, a comparison was made with the naive approaches such as the previous close, exponential moving average (ema) with windows size $p = 5, 10, 15$, and traditional model-based approaches: exponential time-series smoothing (ETS) and autoregressive integrated moving average (ARIMA). The performance was assessed based on the RMSE (root-mean-squared error) $= \sum_{i=1}^{n}(a_i - p_i)^2$ and MAE (mean absolute error) $= \sum_{i=1}^{n}|a_i - p_i|$ for validation and test periods, a_i is the actual value and p_i is the predicted value. The results are first summarised for GSPC in Table 6.4. The integer GA FS proved computationally too burdensome for the BPNN and hence omitted.

Based on the result for GSPC, the following observations were made.

1. SVM and BPNN were able to outperform naive approaches and model- based approaches ARIMA and ETS
2. The reduction in the errors using the SVM was more pronounced compared to the BPNN
3. SVM and BPNN using grammar features selected using integer GA, information gain and mRMR were found to reduce the validation and test errors compared to the same methods using standard technical indicators
4. SVM using integer GA-selected features was able to outperform all other approaches
5. Results were consistent for RMSE and MAE in general

It can be stated that the grammar framework performed well for GSPC in comparison with model-based, naive approaches and ML methods with standard technical

Table 6.4 Performance of different techniques on GSPC

	GSPC result comparison	Validation		Test	
		RMSE	MAE	RMSE	MAE
1	Naive 1—previous close, C_{k-1}	–	–	7.61	5.88
2	Naive 2—ema(C, 5)	–	–	9.39	7.67
3	Naive 3—ema(C, 10)	–	–	11.36	9.49
4	Naive 4—ema(C, 15)	–	–	12.95	10.82
5	ARIMA (1, 1, 1)[a]	–	–	7.58	5.87
6	ETS[a]	–	–	7.60	5.87
BPNN					
	Input features				
7	Standard technical indicators	7.82	6.35	7.86	6.02
9	Grammar—PCA[b]	7.73	6.13	7.69	6.01
10	Grammar—Releif	8.14	6.64	8.05	6.35
11	Grammar—Correlation	7.96	6.37	8.34	6.40
12	Grammar—Info. gain	7.75	6.18	7.69	6.02
13	Grammar—mRMR	7.69	6.07	7.60	5.84
SVM					
	Input features				
14	Standard technical indicators	7.73	6.25	7.54	5.93
15	Grammar—PCA[b]	7.78	6.32	8.10	6.41
16	Grammar—Releif	7.97	6.69	8.03	6.29
17	Grammar—Correlation	8.24	6.14	7.68	6.08
18	Grammar—Info. gain	7.64	6.06	7.51	5.81
19	Grammar—mRMR	7.73	6.08	7.51	5.82
20	Grammar—GA best	**7.63**	**5.98**	**7.45**	**5.80**
21	Grammar—GA average[c]	7.71	6.07	7.49	5.83
22	Grammar—GA worst	7.75	6.11	7.51	5.86

[a] The parameters for the ETS and ARIMA models were chosen using the `forecast` package [10] (see Appendix C C.1 for implementation details)
[b] For fair comparison, only the first 25 principal components were used
[c] Feature subset selection using GA was repeated 10 times with different subset initialisations. The results were averaged over the 10 runs

indicators. It is evident that the generated feature space contains feature combinations with some additional information not captured by the commonly used standard technical indicators. This implies that better solution hypothesis can be formulated by grammar-generated features provided that "good" feature subsets for the particular ML algorithm are selected. FS therefore plays a critical role. Based on the above observations, the results for other indices are compactly presented in Table 6.5. Only the RMSE is reported and the BPNN and SVM performance is evaluated for the standard technical indicators and selected grammar features using mRMR and

Table 6.5 RMSE for validation and test data for major stock indices using the ARIMA, ETS, AR(1), EMA, BPNN using technical indicators (TIs) and grammar feature subsets selected using mRMR and SVM using TIs and grammar feature subsets selected using mRMR and integer GA

Method	HSI		SSMI		SSEC	
	Validation	Test	Validation	Test	Validation	Test
ARIMA	–	138.95	–	47.96	–	19.78
ETS	–	139.95	–	47.13	–	19.78
AR(1)	–	140.24	–	47.60	–	19.71
EMA	–	183.51	–	61.42	–	28.16
BPNN (TIs)	100.00	155.43	34.07	49.22	16.57	20.86
BPNN (Grammar—mRMR)	99.71	140.55	33.56	43.98	16.33	19.02
SVM (TIs)	107.48	136.70	36.04	46.26	16.50	**18.03**
SVM (Grammar—GA Average)	99.71	**135.11**	33.55	**46.17**	15.79	18.54
SVM (Grammar—GA best)	97.30	133.71	33.53	45.73	16.64	18.27
SVM (Grammar—GA worst)	101.33	137.12	33.57	46.55	16.77	18.92
Method	FTSE		N225		NDX	
	Validation	Test	Validation	Test	Validation	Test
ARIMA	–	34.33	–	207.07	–	14.35
ETS	–	33.54	–	207.13	–	14.05
AR(1)	–	33.71	–	207.21	–	**13.98**
EMA	–	42.63	–	275.20	–	14.75
BPNN (TIs)	26.79	36.19	90.09	211.94	15.79	16.90
BPNN (Grammar—mRMR)	25.48	34.61	85.80	221.79	14.14	15.78
SVM (TIs)	26.34	40.27	94.29	203.89	14.89	15.15
SVM (Grammar—GA Average)	25.30	**32.64**	86.49	**203.30**	14.40	14.37
SVM (Grammar—GA Best)	24.77	32.27	85.96	202.70	14.22	14.04
SVM (Grammar—GA Worst)	26.11	33.37	86.49	204.35	15.68	14.75
Method	GDAXI		TWII		AORD	
	Validation	Test	Validation	Test	Validation	Test
ARIMA	–	44.19	–	69.52	–	29.97
ETS	–	44.08	–	69.46	–	29.98
AR(1)	–	44.18	–	69.41	–	29.97
EMA	–	58.26	–	96.55	–	38.81
BPNN (TIs)	31.77	47.73	47.97	71.02	22.43	54.18
BPNN (Grammar—mRMR)	36.69	51.37	46.88	70.94	21.22	73.98
SVM (TIs)	31.04	**43.70**	48.61	**66.85**	26.06	29.23
SVM (Grammar—GA Average)	29.94	44.49	46.99	67.13	21.53	**28.61**
SVM (Grammar—Best)	27.45	43.93	46.39	67.00	21.87	27.95
SVM (Grammar—GA Worst)	30.12	46.28	47.03	67.41	21.04	30.14

integer GA. ARIMA, ETS, AR(1) and EMA = ema(C, p = 5) results are also shown. Based on this empirical study involving a range of stock indices, the following can be stated. The results were consistent when the metric was changed to MAE.

1. In general, the grammar features reduce the validation and test errors in ML. For example, for NDX, the reduction is by (1.52 and 2.82 %) using a BPNN with mRMR and (0.49 and 0.78 %) for the SVM using integer GA. Similarly, for HSI, the best reduction is by (1.80 and 2.57 %) using the BPNN and (7.77 and 0.59 %) using the SVM. The grammar features reduced the validation and test errors for GSPC, HSI, SSMI, SSEC, FTSE, N225, NDX and TWII using the BPNN (8/10 indices) and GSPC, HSI, SSMI, FTSE, N225, NDX and AORD (7/10 indices) using the SVM.
2. The reduction in the errors using the SVM was more pronounced compared to the BPNN. Furthermore, the accuracy of the estimates was higher for the SVM in most cases. The superior performance of SVM might be due to better gener-alisation achieved by structural risk minimisation as opposed to empirical risk minimisation in BPNN [8]. The poor generalisation is clearly observed in BPNN for AORD. Additionally, the large number of free parameters in BPNN (hidden layers, number of hidden nodes, learning rate, momentum, epochs, transfer func-tions and weight initialisation methods) requires extensive empirical parameter estimation to achieve optimal results.
3. For NDX, the model-based approach AR(1) showed superior performance (1/10 indices). The superior performance of model-based approaches in test is most likely due to some structure in the time-series where there is dependence on the history. This is supported by the RMSE which is considerably smaller compared to other indices which seems to suggest that such models provide a reasonable fit to the data. However, it can be observed that the performance of the model-based methods is only marginally better than the ML techniques.
4. In general, integer GA performed the best followed by mRMR and information gain.

The SVM using grammar features was unable to outperform the SVM using standard technical indicators for SSEC, GDAXI and TWII. Although the difference is marginal, this could be due to the following reasons.

(i) The technical indicators have captured most of the information; hence, the grammar features do not add any additional information
(ii) The FS criteria have failed to select *good* features that are still *good* in test period
(iii) The parameter tuning is far from optimal

Considering that the approach offered an improvement for all other indices, the most probable reason is that the FS criteria have selected less informative features. The time-series is highly non-stationary; hence, the *goodness* of features selected using the validation period might have changed considerably in the test period. It is likely that a selected subset will not be optimal over the entire validation and test

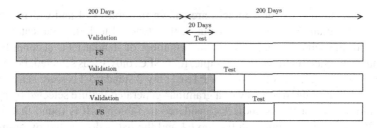

Fig. 6.1 Adaptive FS using the sliding window technique

interval. Based on the belief that the features might have been outdated, the same experiment was conducted again but with adaptive FS once every 20 days as opposed to one-time FS using a validation period of 200 days. The adaptive FS process is depicted in Fig. 6.1. The test period was kept the same and adaptive FS was done by using integer GA as the FS technique. Only the SVM was used.

Furthermore, the best performance for NDX using ML techniques was 14.37 using SVM with integer GA-based FS which is poor compared to the AR(1)/ ETS results of 13.98/14.05. Therefore, adaptive FS was performed on SSEC, GDAXI, TWII and NDX to see whether the results can be improved. The results for the 4 selected indices are brought up in Table 6.6.

It is seen that the errors can be improved by adaptive FS. The SSEC test error in fact outperformed the standard TI indicators (making 8/10 indices in favour of grammar features). Adaptive FS in this manner is not possible if a fixed set of standard technical indicators and parameter values are used as features. It is expected that by repeating feature subset selection as often as necessary, the accuracy of the ML techniques can be further improved. The interval over which selection needs to be repeated will need to be empirically decided. Although this technique is time-consuming and computationally expensive, real-world applications can exploit cloud resources for fast calculations.

As a by-product of the empirical study, potential candidates for technical indicators that consistently improve validation and test errors compared to the standard technical indicators can be identified. For a given stock index, FS using the GA method was repeated 10 times with different initialisation and a histogram was

Table 6.6 RMSE for test period using ARIMA, ETS and SVM with features as TIs, grammar features selected only once and adaptive grammar FS using integer GA

Index	ETS	ARIMA	TIs	Grammar	Adaptive grammar
SSEC	19.78	19.78	18.03	18.54	**17.96**
GDAXI	44.08	44.19	**43.70**	44.49	44.45
TWII	69.46	69.52	**66.85**	67.13	67.88
NDX	**14.05**	14.35	15.15	14.37	14.30

constructed. The results for 4 indices are shown in Tables C.4 and C.5 in Appendix C. Of the grammar-generated features, the standard technical indicators (TIs) are in bold. In each case, only a very small number of standard TIs account for the grammar-generated features. For example, for GSPC, only 1 was selected. For SSMI, no standard TIs were selected, and for FTSE, only 2 standard TIs were selected.

Table 6.7 shows the frequency of a grammar-generated feature selected by aggregating the results for the 10 indices considered in Table 6.1. It is found that only 7 of the 25 standard TIs appear in the top 64 and the list is dominated by grammar-generated features. The actual n and k values are not shown so that the general formulae can be seen. The other frequently selected TIs in order were K (9), ROC (8), OSCP (8), Slow K (8), SMA (8), ATR (6), R (6), ADO (4) and Chaikin volatility (4). In this manner, appropriate parameterised feature combinations can be identified for the time-series under consideration for a particular algorithm.

6.2 Predicting Peak Electricity Load Data

The EUNITE dataset [11] has been extensively used as a benchmark to test load prediction algorithms. It consists of half-hourly recordings in the years 1997 and 1998, and also holiday and temperature information. It was originally published for a competition to predict the peak daily load for January 1999 using half-hourly recordings in the years 1997 and 1998. The performance metric used in electricity load time-series prediction, mean absolute percentage error (MAPE) $= \frac{100}{n} \sum_{i=1}^{n} \left| \frac{y_i - \hat{y}_i}{y_i} \right|$ where $y_i, \hat{y}_i$ are the actual and predicted values respectively was used for evaluation.

Calendar information was encoded as 7 binary values to represent the day of week and a single binary value for holidays as the competition winners in [12]. A self-organising map (SOM) was used to identify data clusters using a peak load history window of 7 days, temperature and calendar information as SOM inputs. This showed a very strong seasonality, dividing the data into *cold* and *hot* seasons (season 1 and 2 in Fig. 6.2, respectively). Based on the SOM result, only the data from season 1 months were used to construct models in the experiments, e.g. to predict Jan. 1999, the model was constructed using January 1997–March 1997, October 1997–March 1998 and October 1998–December 1998 data.

Two different experiments were performed on the dataset. In the first trial, the wavelet-based grammar in Table 4.3 was used in a month-ahead manner of prediction. Here, after predicting a load value for January 1st, 1999, this predicted value was used with the historical values before January 1st to predict January 2nd, 1999. This process was continued until the predicted load value of January 31st was obtained. For the second approach, the grammar in Table 4.10 was used. This required the previous day's high, low and close electricity loads (HLC), and was suited to a day-ahead prediction approach. Therefore, the actual historical HLC values were used in predicting each day. The usage of HLC values in electricity load prediction is unconventional but conceptually justified because the data are better represented.

Table 6.7 Technical indicators and selected grammar feature frequency

	Feature	Frequency
1	**Disparity**	32
2	`C-(sd(lag(C, k), n))`	24
3	`C-(sd(lag(M, k), n))`	23
4	`C-(sd(lag(L, k), n))`	22
5	`H-(sd(lag(C, k), n))`	21
6	`L-(sd(lag(L, k), n))`	21
7	`sd(diff(H), n))/(sd(delt(L), n))`	21
8	`(C-(M-lag(C, k)))/n`	20
9	`(C-(M-lag(C, k)))/n`	19
10	**Bias**	18
11	`L-(sd(lag(H, k), n))`	18
12	`(M-(H-lag(H, k)))/n`	18
13	`(C-(M-lag(H, k)))/n`	17
14	`H-(sd(lag(L, k), n))`	17
15	`L-(sd(lag(M, k), n))`	17
16	`(L-(H-lag(L, k)))/n`	17
17	`(sd(diff(H), n))/(sd(delt(M), n))`	17
18	**Lower bollinger band**	17
19	`(C-(meandev(M, n)))/n`	16
20	`(H-(L-lag(H, k)))/n`	16
21	`M-(sd(lag(M, k), n))`	16
22	`M-(sma(ema(diff(L), n), n))`	16
23	`(C-(H-lag(C, k)))/n`	15
24	`C-(sd(diff(C), n))`	15
25	`C-(sma(ema(diff(L), n), n))`	15
26	`L-(sd(diff(C), n))`	15
27	`sma(L, n) - 2*(sd(L, n))`	15
28	`(C-(H-lag(H, k)))/n`	14
29	`(C-(L-lag(C, k)))/n`	14
30	`(C-(L-lag(H, k)))/n`	14
31	`(C-(L-sma(L, n)))/n`	14
32	**CLV**	14
33	`(lag(M, k)) - (sd(lag(H, k), n))`	14
34	`M-(sma(ema(diff(M), n), n))`	14
35	`sma(L, n) + 2*(sd(L, n))`	14
36	`C-(H-lag(L, k)))/n`	13
37	`C-(H-sma(L, n)))/n`	13
38	`C-(sd(lag(H, k), n))`	13
39	`H-(sd(lag(M, k), n))`	13

(continued)

Table 6.7 (continued)

	Feature	Frequency
40	`(M-(L-lag(H, k)))/n`	13
41	`(sd(diff(C), n))/(sd(delt(C), n))`	13
42	`(sd(diff(C), n))/(sd(delt(M), n))`	13
43	`(sd(diff(L), n))/(sd(delt(M), n))`	13
44	`sma(M, n) + 2*(sd(M, n))`	13
45	`C-(sma(diff(H), n))`	12
46	`C-(sma(ema(diff(M), n), n))`	12
47	`H-(sma(diff(H), n))`	12
48	`L-(sd(lag(C, k), n))`	12
49	`M-(sd(lag(L, k), n))`	12
50	`(L-(H-lag(C, k)))/n`	12
51	`(M-(L-lag(C, k)))/n`	12
52	`(sd(diff(M), n))/(sd(delt(L), n))`	12
53	**Upper Bollinger Band**	12
54	**Aroon**	11
55	`(C-(H-lag(M, k)))/n`	11
56	`(C-(M-lag(L, k)))/n`	11
57	`(H-(meandev(L, n)))/n`	11
58	**Lagged closing price**	11
59	`H-(sd(diff(L), n))`	11
60	`H-(sd(lag(H, k), n))`	11
61	`M-(sd(lag(C, k), n))`	11
62	`M-(sma(ema(diff(H), n), n))`	11
63	`(sum(L, n))/(max(i^+,n))`	11
64	`sma(H, n) + 2*(sd(H, n))`	11

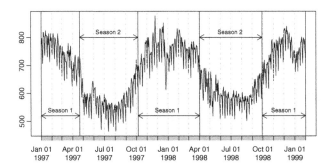

Fig. 6.2 Clustering days to different seasons using a SOM

For initial performance comparison, ARIMA and ETS models were chosen as analytical methods. SVM and kernel recursive least squares (KRLS) algorithm were used as kernel-based ML methods. Table 6.8 summarises the results of the different techniques used for different feature subsets on different months. It was observed that the best results were produced by a radial basis function (RBF) kernel.

Based on the features used in previous load prediction work, the performance of 5 domain-expert suggested feature subsets was compared in Table 6.8. It was observed that for a particular feature subset, the result stability for different months is poor, e.g. the feature subset $H_k, \Delta H_k, \Delta D_i^1, \Delta D_i^2, \Delta D_i^3, \Delta S_i (\forall i \in \{k-1, ..., k-6\})$ performed remarkably well only in January 1999, but the average accuracy was inferior to other subsets considered. It can be suspected that some related works on the EUNITE dataset have used specific subsets in a trial and error fashion to report good performance only on the *test* dataset, January 1999.

Grammatical evolution (GE) was used as the FS method as described in Sect. 5.3.3. GE has no selective feature pruning, and the GE fitness function is crafted to evaluate feature subset goodness. As KRLS requires one less parameter (σ and λ) than support vector regression (C, γ and ε) consuming less hyper-parameter tuning time, KRLS was chosen for feature subset evaluation as explained in Algorithm 4 Sect. 5.3.3. The optimal parameters for the KRLS feature subset evaluator, $\sigma = 256$ and $\lambda = 0.0625$ were chosen by ten-fold time-series cross validation using a peak load history window of 7 days and encoded calendar information as features. A chromosome of the proposed GE algorithm comprised of 24 codons and 25 genes, i.e. a maximum of 25 features per subset with each feature represented by 24 codons making the integer chromosome size 600. The population size was 48, and 100 generations were iterated. Roulette wheel parent selection technique was used and the proposed multi-point crossover technique in Fig. 4.7 was used. The best 2 chromosomes in each generation were considered to be elite individuals.

A total of 10 trials were performed and the results are presented at the bottom of Table 6.8. The HLC-based grammar produced significantly better results for each of the 3 months (1.15, 1.73 and 1.39 %), and even the worst performing HLC feature subset outperformed all other methods for average day-ahead predictions. Although the best wavelet feature subset (1.81 %) outperformed other domain-expert-suggested feature subsets for day-ahead predictions, the average performance of each month (1.96 %) was poor. However, its performance for month-ahead prediction was promising. This leads to state that the HLC grammar incorporates more information about the daily variations of load and significantly improves day-ahead predictions, while the wavelet grammar elucidates long-term trends and hence is more suited for month-ahead predictions. The best one-day-ahead predictions are plotted against the actual values and the cumulative MAPE is also shown in Fig. 6.3.

Table 6.9 compiles selected features from all GE trials filtered using the maximum-relevance-minimum-redundancy (mRMR) criterion. While some features are obvious and straightforward (e.g. ΔH, H, delt(H), histwin(H,14)), many others are not so apparent to human experts.

The best results achieved in the literature, summarised in Table 6.10, all use a window of past days' peak load, temperature and calendar information. The main

Table 6.8 Kernel method performance comparison for different feature subsets on 3 different months

Method and features	Month-ahead MAPE %				Day-ahead MAPE %			
	January 1999	December 1998	November 1998	Average	January 1999	December 1998	November 1998	Average
Last year's data	2.29	4.79	3.22	3.43	2.29	4.79	3.22	3.43
ARIMA	2.08	4.97	3.45	3.50	2.60	2.55	2.71	2.62
ETS	1.87	4.66	3.47	3.33	2.11	2.48	2.10	2.23
Linear SVM with $H_i(\forall i \in \{k,...,k-6\})$	2.32	3.84	1.82	2.66	1.62	2.39	1.99	2.00
Polynomial SVM with $H_i(\forall i \in \{k,...,k-6\})$	2.22	3.50	1.82	2.52	1.72	2.39	1.99	2.03
Radial SVM with $H_i(\forall i \in \{k,...,k-6\})$	1.96	2.87	1.75	2.19	1.68	2.19	1.81	**1.89**
Polynomial KRLS with $H_i(\forall i \in \{k,...,k-6\})$	3.93	4.71	3.46	4.03	3.91	4.62	3.38	3.97
Radial KRLS with $H_i(\forall i \in \{k,...,k-6\})$	1.67	3.06	1.76	**2.16**	1.61	2.31	1.85	1.92
Using radial KRLS with feature subsets								
$H_i(\forall i \in \{k,...,k-6\})$ + Temp.	3.59	3.05	2.06	2.90	2.25	2.52	1.86	2.21
$H_i(\forall i \in \{k,...,k-6\})$	1.67	3.06	1.76	**2.16**	1.61	2.31	1.85	1.92
D_k^1, D_k^2, D_k^3, S_k	1.88	3.25	1.70	2.27	1.62	2.05	1.84	**1.84**
$H_k, \Delta H_k, \Delta D_k^1, \Delta D_k^2, \Delta D_k^3, \Delta S_k$	1.64	3.49	1.96	2.36	1.62	2.08	1.93	1.88
$H_k, \Delta H_k, \Delta D_i^1, \Delta D_i^2, \Delta D_i^3, \Delta S_i (\forall i \in \{k-1,...,k-6\})$	1.55	3.98	2.56	2.70	1.25	2.61	2.25	2.04

(continued)

Table 6.8 (continued)

Method and features	Month-ahead MAPE %				Day-ahead MAPE %			
	January 1999	December 1998	November 1998	Average	January 1999	December 1998	November 1998	Average
HLC grammar—best of each month	–	–	–	–	1.15	1.73	1.39	**1.48**
HLC grammar—average of each month	–	–	–	–	1.34	1.89	1.56	1.60
HLC grammar—worst of each month	–	–	–	–	1.62	2.03	1.68	1.68
Wavelet grammar—best of each month	1.42	2.45	1.35	**1.84**	1.61	1.93	1.66	1.81
Wavelet grammar—average of each month	1.82	2.89	1.54	2.08	1.77	2.31	1.79	1.96
Wavelet grammar—worst of each month	2.42	3.32	1.76	2.31	1.90	2.89	1.98	2.26

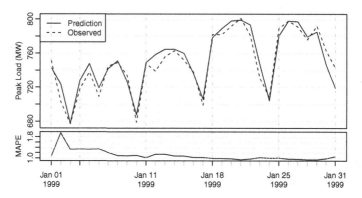

Fig. 6.3 The best one-day-ahead prediction values versus the actual values (MAPE 1.15%)

Table 6.9 Selected features from 10 GE runs

	HLC features	Wavelet features
1	`(|C-L|-|L|)/|C-L|`	D_1
2	`(|L|-|H-L|)/L`	`delt(H)`
3	`(ema(ΔC, 2))/(ema(delt(C), 2))`	ΔD_1
4	`(ema(H, 5))/(min(ema(H, 2), 2))`	ΔH
5	`delt(H)`	ΔS
6	`delt(L)`	`H`
7	ΔH	`lag(D_3,5)`
8	`ema(C, 3)`	`lag(delt(H), 2)`
9	`ema(C-L, 5)+sd(H-L, 3)`	`lag(ΔD_1, 1)`
10	`ema(D, 3)`	`lag(ΔD_3, 4)`
11	`ema(ema(H, 5), 3)`	`lag(meandev(D_1, 14), 3)`
12	`ema(H, 7)`	`lag(sd(H, 7), 3)`
13	`ema(H, 7)-sd(H-C, 2)`	`lag(S, 6)`
14	`ema(H-C, 2)+sd(H, 14)`	`lag(sma(D_1, 7), 6)`
15	`ema(lag(D,7), 3)`	`histwin(ΔH, 7)`
16	`ema(lag(U, 3), 3)`	`lag(sma(H, 5), 4)`
17	`lag(H, 7)`	`histwin(H, 14)`
18	`lag(U, 1)`	`meandev(D_1,14)`
19	`max(H, 14)`	`sd(D_1, 14)`
20	`max(L, 3)-lag(H, 7)`	`sd(H, 14)`
21	`max(lag(H, 1), 3)`	`S`
22	`min(D, 5)`	`sma(D_1, 14)`
23	`min(ΔC, 5)`	`sma(D_1,7)`
24	`sd(H, 5)-lag(D, 7)`	`sma(D_2,3)`
25	`sma(ΔL, 2)`	`sma(D_3,3)`

Table 6.10 Benchmark results on the EUNITE dataset (Compiled from [13, 14])

Method	MAPE (%)	Year
SVM (EUNITE competition winner)	1.95	2004
SVM-GA	1.93	2006
Autonomous ANN	1.75	2007
Floating search + SVM	1.70	2004
MLP-NN + Levenberg-Marquardt	1.60	2008
Auto-regressive recurrent ANN	1.57	2006
Local prediction framework + SVR	1.52	2009
Feedforward ANN	1.42	2005
Best GE result	**1.42**	–
SOFNN + Bi-level optimisation	1.40	2009
LS-SVM + chaos theory	1.10	2006
MLP-NN + Differential evolution	1.02	2011

differences between the different methods lie in the data partitioning technique, choice of the ML algorithm and its training method. There is some ambiguity in these results in that in the original competition, the temperature results of the testing period as well as the actual load data was not available. After the competition, some researchers might have used this information in order to improve their results and the original competition winner's result has been *claimed* to be outperformed. Furthermore, some works do not state if it is day-ahead or month-ahead predictions. The results discussed in this section were also published in [15].

6.3 Predicting Foreign Exchange Client Trade Volume

The foreign exchange market enables global firms to easily convert currencies crucial for trade and investment. Banks are an important part of this market, brokering for small institutes and big corporations with minimal costs thus creating specialised services for international traders. With the floating nature of exchange rates, traders and brokers are exposed to financial risk when trading this market. Exchange rate fluctuations can cause profit and loss to both brokers and traders alike. While speculative traders are attracted to this risk, corporations hedge to reduce their risk. Different financial instruments are created to reduce risk exposure of financial bodies, but to remain competitive in the already efficient foreign exchange market, banks are seeking to exploit machine intelligence to better hedge their risk. Given the exchange rate and customers history, can the risk exposure of a bank be reduced using ML techniques? The literature on industrial bank/broker hedging is scarce due to the proprietary nature of in-house-developed techniques. In a hedging system, the prediction engine is a major component. This section explains the implementation of a grammar-based prediction engine for foreign exchange client trade volume prediction.

Data Preprocessing: The dataset contained transaction details for different currency pairs. The majority of the transactions were AUD/USD. These were spot transactions and the data were irregular. The AUD/USD transactions were aggregated to form hourly net trade volume time-series V. Firstly, net trade volume prediction task was formulated as a binary classification problem. $V > 0$ meant that the net trade volume was "buy" and vice versa. The out-sample test results for 6 months are presented in Appendix C Table C.3. The first observation is that the data are unbalanced (1,460 net buys and 604 net sells) for the 6 months considered. Therefore, the balanced accuracy is a more suitable measure to assess the model performance and the 6 month average is 51.06 %. It was then attempted to deal with a more balanced dataset, and a threshold *Thresh* was decided to construct classification labels as,

$$
class = \begin{cases} -1 & V < -Thresh \ (1{,}085 \text{ samples}) \\ 0 & -Thresh < V < Thresh \ (959 \text{ samples}) \\ +1 & V > Thresh \ (660 \text{ samples}) \end{cases}
$$

Candidate Features: Unlike stock market index and electricity load data, previous knowledge of the types of features to use was unavailable. Therefore, after manually picking features, the best results were achieved by using the following features.

1. 6 external features that can affect the trading behaviour of small- and medium-scale foreign exchange traders, e.g. T-bill rates, interest rates, bond prices, first difference of the AUD/USD foreign exchange rate obtained from SIRCA [16]
2. A client trading volume history window of 16 h, i.e. lag(V, 1), ..., lag(V, 16)
3. Time-derived features: Month of the year (1–12), day of the week (1–7), day of the month (1–31) and hour of day (0–23)
4. First difference of the trade volume wavelets, $\Delta(D1)$, $\Delta(D2)$, $\Delta(D3)$ and $\Delta(S)$

The grammar features considered with varying lags k and moving window lengths and look-back periods n for predicting foreign exchange client trade volume were,

 (i) Wavelet features: Grammar features derived from the 3-level decomposition of the client trade volume time-series for different n, k
 (ii) Moving average features: Grammar features derived from a combination of moving averages applied on the client trade volume time-series for different n
(iii) Lag-based features: Grammar features derived using different lags, maximum and minimum values for different n, k values of the client trade volume time-series
(iv) Exchange rate-related features: Technical indicator type grammar features using the exchange rate which can be believed to be a driving factor of client trades

From each of the 4 grammar families, 10 features were first selected using mRMR to create a pool of 40 features and the binary GA wrapper was used to select feature subsets from this pool as in Fig. 6.4. The resultant best chromosome encoding represented the selected features and the number of '1's in the chromosome was the

Fig. 6.4 Features considered for client trade volume prediction. Grammar features were supplementary to the manually selected features

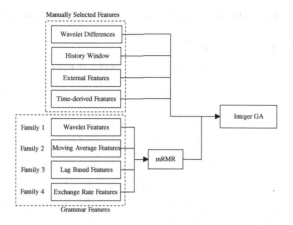

feature subset size. Adaptive FS was performed weekly. SVM was used with a RBF kernel and the parameters were also re-optimised weekly. The predictions were done in a rolling-window manner to get day-ahead predictions. The SVM parameters were optimised (Table 6.11).

Confusion Matrix: Columns represent predicted class and rows represent actual class, e.g. for a binary classification task, $\begin{pmatrix} 15 & 33 \\ 56 & 45 \end{pmatrix}$, TN = 15, FP = 33, FN = 56 and TP = 45.

The out-sample test results for 8 months are presented in Table 6.12 and for 6 out of 8 months, the grammar improved the 3-class classification accuracies. The overall balanced accuracy improved from 38.03 to 38.91 % and the overall hit ratio improved from 42.05 to 42.86 %. A closer inspection of the confusion matrix showed that the class −1 prediction accuracy remained the same at 696/1085, while class 0 accuracy improved from 357/959 to 372/959 and class +1 accuracy improved from 84/660 to 91/660.

The grammar-based prediction system also enabled the identification of the best features to use with the selected ML algorithms. The parameters of the moving averages, look-back periods of maximum, minimum values and the appropriate lags to use can also be empirically decided in this manner, e.g. a window size of 4 was associated with a large number of features which means that recent history is more important. The best features discovered for the optimised SVM are in Table 6.11. V is the net trade volume and i^+, i^- are the days elapsed since the last highest net trade volume and the last lowest net trade volume. F is the AUD/USD foreign exchange rate.

An interesting observation is the absence of wavelet-based grammar features on the top 30 features. Therefore, when attempting to predict this type of data, the effectiveness of wavelets should be further investigated. Net trade volume lag-based features were dominant and the exchange rate seemed to influence trade volumes as well.

Table 6.11 Selected grammar feature frequency for different families

	Feature	Frequency	Family
1	$i^-(4)$	13	3
2	$\mathtt{sma}(\Delta(\mathtt{F}),16) - \mathtt{ema}(\Delta(\mathtt{F}),4)$	11	4
3	$(i^-(8) - i^-(4))/4$	11	3
4	$i^+(4)$	10	3
5	$i^-(4)/(\mathtt{lag}(\min(\mathtt{V},16),\ 16)$	9	3
6	$i^+(16)/(\mathtt{lag}(\min(\mathtt{V},16),\ 4)$	9	3
7	$(i^-(16)\text{-}i^-(4))/4$	8	3
8	$\mathtt{ema}(\mathtt{lag}(\mathtt{F},8),4)/\mathtt{sma}(\mathtt{lag}(\mathtt{F},8),4)$	8	4
9	$\mathtt{sma}(\Delta(\mathtt{V}),32)/\mathtt{ema}(\Delta(\mathtt{V}),4)$	8	2
10	$\mathtt{sma}(\Delta(\mathtt{F}),32)/\mathtt{sma}(\Delta(\mathtt{F}),4)$	8	4
11	$\mathtt{wma}(\Delta(\mathtt{F}),4)/\mathtt{ema}(\Delta(\mathtt{F}),4)$	8	4
12	$(i^-(16)\ -\ i^-(8))/4$	8	3
13	$(i^+(4)\ -\ i^-(8))/4$	8	3
14	$(\min(\mathtt{V},16)\text{-}\min(\mathtt{V},8))/4$	8	3
15	$(\min(\mathtt{V},\ 4)\text{-}\mathtt{V}))/4$	7	3
16	$(\min(\mathtt{V},8)\text{-}\min(\mathtt{V},4))/4$	7	3
17	$i^+(16)/(\mathtt{lag}(\min(\mathtt{V},16),\ 16)$	6	3
18	$i^-(8)$	6	3
19	$\mathtt{V}/\min(\mathtt{V},16)$	6	3
20	$(\min(\mathtt{V},16)\text{-}\min(\mathtt{V},4))/4$	5	3
21	$((\min(\mathtt{V},16)\text{-}\mathtt{V}))/4$	5	3
22	$i^-(4)/(\mathtt{lag}(\min(\mathtt{V},16),\ 4)$	4	3
23	$i^-(8)/(\mathtt{lag}(\min(\mathtt{V},16),\ 16)$	4	3
24	$i^+(8)/(\min(\mathtt{V},16)$	4	3
25	$(i^-(16)\text{-}i^-(8))/4$	4	3
26	$\Delta(\mathtt{V})\ -\ \mathtt{sma}(\Delta(\mathtt{V}),8)$	4	2
27	$\mathtt{ema}(\Delta(\mathtt{V}),16)\ -\ \mathtt{wma}(\Delta(\mathtt{V}),16)$	4	2
28	$\mathtt{sma}(\Delta(\mathtt{V}),16)/\mathtt{ema}(\Delta(\mathtt{V}),16)$	4	2
29	$i^-(16)$	4	2
30	$(\mathtt{wma}(|V|,\ 4))/(\mathtt{wma}(\Delta(\mathtt{V}),\ 4))$	4	2

Table 6.12 Out-sample results (%) using binary GA to predict the client trade volume classification for 8 months using SVM

Month	Method	Confusion matrix			Balanced accuracy	Hit ratio
1	Without grammar	84	41	16	35.69	38.64
		55	41	17		
		56	31	11		
	With grammar	91	31	19	**37.96**	**40.91**
		61	35	17		
		50	30	18		
2	Without grammar	80	31	18	37.11	39.29
		57	33	29		
		60	9	19		
	With grammar	77	36	16	**38.38**	**40.77**
		57	43	19		
		56	15	17		
3	Without grammar	65	63	2	**37.79**	**43.47**
		53	87	0		
		31	50	1		
	With grammar	63	62	6	37.27	42.90
		52	87	1		
		28	53	1		
4	Without grammar	168	4	0	35.17	**52.38**
		94	8	0		
		57	5	0		
	With grammar	158	14	0	**35.85**	51.79
		86	16	0		
		57	5	0		
5	Without grammar	100	32	6	36.63	41.37
		68	31	10		
		62	19	8		
	With grammar	97	29	12	**37.40**	**41.67**
		64	31	14		
		57	20	12		
6	Without grammar	94	11	30	**36.32**	**38.64**
		62	11	39		
		67	7	31		
	With grammar	84	18	33	34.79	36.65
		63	12	37		
		57	15	33		
7	Without grammar	59	60	14	32.42	37.20
		64	60	13		
		40	20	6		
	With grammar	64	56	13	**35.35**	**41.07**
		59	69	9		
		40	21	5		

(continued)

Table 6.12   (continued)

Month	Method	Confusion matrix			Balanced accuracy	Hit ratio
8	Without grammar	46	58	3	40.71	46.05
		31	86	10		
		26	36	8		
	With grammar	62	38	7	**42.43**	**58.03**
		39	79	9		
		40	25	5		
1–8	Without grammar	696	300	89	38.03	42.05
		484	357	118		
		399	177	84		
	With grammar	696	284	105	**38.91**	**42.86**
		481	372	106		
		385	184	91		

References

1. J.A. Ryan, Quantmod: Quantitative financial modelling framework. (2013), http://CRAN.R-project.org/package=quantmod. R package version 0.4-0
2. J. Knaus, Snowfall: Easier cluster computing (based on snow). (2010), http://CRAN.R-project.org/package=snowfall. R package version 1.84
3. D. Meyer, E. Dimitriadou, K. Hornik, A. Weingessel, e1071: Misc functions of the department of statistics. (TU Wien, 2012), http://CRAN.R-project.org/package=e1071. R package version 1.6-1
4. A. Lendasse, E. de Bodt, V. Wertz, M. Verleysen, Non-linear financial time series forecasting - Application to the BEL-20 stock market index. Eur. J. Econ. Soc. Syst. **14**(1), 81–91 (2000)
5. J. Kamruzzaman, R.A. Sarker, Forecasting of currency exchange rates using ANN: A case study, in *Proceedings of the 2003 International Conference on Neural Networks and Signal Processing*, vol. 1 (2003), pp. 793–797
6. K.-J. Kim, Financial time series forecasting using support vector machines. Neurocomputing **55**(1), 307–319 (2003)
7. W. Shen, X. Guo, Forecasting stock indices using radial basis function neural networks optimized by artificial fish swarm algorithm. Knowl.-Based Syst. **24**(3), 378–385 (2011)
8. F.E.H. Tay, L. Cao, Application of support vector machines in financial time series forecasting. Omega **29**(4), 309–317 (2001)
9. A. Zapranis, Testing the random walk hypothesis with neural networks, *Artificial Neural Networks ICANN 2006*, vol. 4132 of Lecture Notes in Computer Science (Springer, Heidelberg, 2006), pp. 664–671
10. R.J. Hyndman et al., Forecast: Forecasting functions for time series and linear models. (2013), http://CRAN.R-project.org/package=forecast. R package version 4.06
11. EUNITE, World-wide competition within the EUNITE network. (2001), http://neuron-ai.tuke.sk/competition/
12. B.-J. Chen, M.-W. Chang et al., Load forecasting using support vector machines: A study on EUNITE competition 2001. IEEE Trans. Power Syst. **19**(4), 1821–1830 (2004)
13. M. Moazzami, A. Khodabakhshian, R. Hooshmand, A new hybrid day-ahead peak load forecasting method for Iranian national grid. Appl. Energy **101**, 489–501 (2013)
14. J. Nagi, F. Nagi, S.K. Tiong, S.K. Ahmed, K.S. Yap, A computational intelligence scheme for the prediction of the daily peak load. Appl. Soft Comput. **11**(8), 4773–4788 (2011)

15. A.M. De Silva, F. Noorian, R.I.A. Davis, P.H.W. Leong, A hybrid feature selection and generation algorithm for electricity load prediction using grammatical evolution, in *ICMLA* (2). (2013), pp. 211–217
16. Sirca, Enabling financial research (October 2013), www.sirca.org.au/

Chapter 7
Conclusion

The performance of machine learning (ML) techniques is highly dependant on the formalism in which the solution hypothesis is represented. The features used to formalise this hypothesis should be engineered carefully for optimal performance. This is usually done by domain experts which often leads to good results. This brief investigated if an automatic feature generation framework that can generate expert suggested features and many other parametrized features can be used to improve the performance of ML methods in time-series prediction.

The feature generation framework using context-free grammars (CFGs) was proposed in Chap. 4. A key benefit is that it enables experts to provide guidelines to the system but for the computer to search for a good solution among a large number of candidates. Grammar families were proposed as a way to organise and constrain the feature space. Pruning strategies that can be used to eliminate features without compromising the effectiveness of the feature generation were discussed. The merits of the proposed framework are, (i) parametrized features can be engineered in a systematic manner based on time-series dynamics (non ad-hoc parametrization) (ii) the user is able to design a sufficiently large grammar to capture as much information as possible with a manageable feature space (iii) the user can incorporate domain knowledge by choosing appropriate derived variables and production rules. It was realised that not only the expansion but mining the generated feature space to discover better feature subsets is also important.

Chapter 5 described how different feature selection (FS) criteria were explored to mine the expanded feature space. The wrapper-based techniques were found to work better than filter-based approaches. Integer genetic algorithms, maximum-relevance-minimal-redundancy and information gain criteria performed the best.

© The Author(s) 2015
A.M. De Silva and P.H.W. Leong, *Grammar-Based Feature Generation*
for Time-Series Prediction, SpringerBriefs in Computational Intelligence,
DOI 10.1007/978-981-287-411-5_7

Grammatical evolution was proposed as a hybrid feature generation and selection algorithm. This hybrid algorithm combines the feature generation and selection phases and avoids the need for selective feature pruning strategies. The fitness function of the genetic algorithm was crafted to penalise *bad* feature subsets and select *good* feature subsets using a wrapper system.

The empirical evaluation in Chap. 6 using real-world financial and electricity load time-series data showed that the proposed approach can lead to improvements in performance for particular ML algorithms that use certain expert suggested feature subsets. These suggested features for financial time-series prediction were chosen as widely used standard technical indicators. For electricity load time-series, different features used in previous works such as history windows, wavelet transforms and differenced values were chosen.

The classification accuracy of foreign exchange client trade volume time-series was also improved by using grammar families. Because of the absence of literature on predicting such time-series, features engineered manually and exogenous features were defined as expert features.

Although the proposed approach is not *guaranteed* to produce better out-sample results, it can certainly be used as a technique to craft features. The best approach is to begin with expert suggested features and use the system generated feature combinations as supplementary feature subsets to add value to the predictor. An experienced user can monitor features that consistently rank in the top and consider them as *good* features. The feature parameters and the optimal feature subsets are subject to change due to the non-stationary nature of real-world time-series hence it is intuitive that an expert system can assist a human expert in selecting the best feature combinations automatically.

In a nutshell, the proposed framework is a systematic approach to generate a rich class of features from an expert designed set of rules that can yield considerable improvement in the performance of any ML technique in a practical sense for critical applications.

Future Work The grammar developed in this work was designed to generate only a limited feature space. For example, the grammar for stock market index time-series generated only a limited set of technical indicators. Advanced conditional technical indicators can be generated by combining probabilistic context-free grammar and genetic programming. The grammar can also be applied to other transformations of the time-series, e.g. the grammar for electricity load time-series can have other effective transformations such as empirical mode decomposition (EMD).

In addition, robust FS for non-stationary time-series using ensemble approaches and non-linear dimension reduction techniques can also be investigated further. It was discovered that wrapper approaches are better than filter approaches hence advanced FS algorithms should be developed that will ensure that the selected feature subsets are robust for non-stationary time-series which will aid in minimising the gap between the best and worst performing feature subsets, e.g. in Table 6.8.

It was also discovered that adaptive FS can lead to better results. This can be supplemented with adaptive parameter optimisation to account for the non-stationary

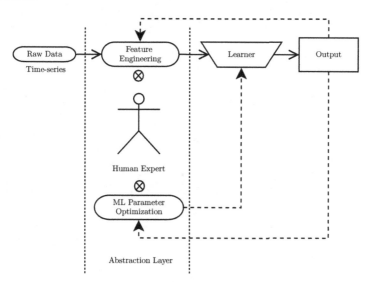

Fig. 7.1 The proposed system viewed as an abstraction layer

nature of time-series. Although this is time-consuming, small improvements can yield benefits in critical application such as financial time-series prediction.

The proposed approach is general and is not limited to time-series investigated in the brief. Other time-series can be explored and appropriate grammars can be developed. By using the proposed approach, an insight to the feature combinations that work well with particular ML algorithms can be discovered hence experts can use the proposed feature generation framework in any application to supplement their own set of hand picked features.

The approach is computationally expensive hence certain experiment parameters were restricted, e.g. number of generations in genetic algorithm, tuning parameter grid size, etc. A cloud-based parallel implementation can speed up the feature generation and selection process in real-world applications.

Finally, as in Fig.7.1, the approach can be viewed as an abstraction layer to reduce the effort associated with finding the best feature-ML architecture-parameter combination, the most time-consuming aspect in developing a ML system, while simultaneously enhancing performance.

Appendix A
Wavelet Transformations

A.1 Discrete Wavelet Transform (DWT)

This appendix is loosely based on the presentation outline in a technical report by Saif et al.[1]

DWT decomposes a time-series into a set of orthonormal basis functions (wavelets) producing a set of wavelet coefficients. Each coefficients captures information at different frequencies at distinct times. A function f can be expanded using a mother wavelet function Ψ as,

$$f(t) = \sum_{j=-\infty}^{\infty} \sum_{k=-\infty}^{\infty} w_{jk} 2^{j/2} \Psi(2^j t - k) \qquad (A.1)$$

The functions $\Psi(2^j t - k)$ are mutually orthogonal. The coefficient w_{jk} conveys information about the behaviour of f at a scale around 2^{-j} near time $k \times (2^{-j})$. Application of DWT for time-series analysis suffers from a lack of translation invariance which is undesired. The highly redundant maximal overlap discrete wavelet transform (MODWT) is used to address this issue.

A.2 Maximal Overlap DWT (MODWT) and À Trous Filtering

When MODWT is applied on an N sample input time-series it will produce N samples for each resolution level aligning with the original time-series in a meaningful way. For a time-series X with N samples, the jth level MODWT wavelet ($\tilde{W}_j$) and scaling ($\tilde{V}_j$) coefficients are,

---

[1] Saif Ahmad, Ademola Popoola, and Khurshid Ahmad. Wavelet-based multiresolution forecasting. *University of Surrey, Technical Report*, 2005.

© The Author(s) 2015
A.M. De Silva and P.H.W. Leong, *Grammar-Based Feature Generation for Time-Series Prediction*, SpringerBriefs in Computational Intelligence, DOI 10.1007/978-981-287-411-5

$$\tilde{W}_{j,t} = \sum_{l=0}^{L_{j-1}} \tilde{h}_{j,l} X_{t-l \bmod N} \tag{A.2}$$

$$\tilde{V}_{j,t} = \sum_{l=0}^{L_{j-1}} \tilde{g}_{j,l} X_{t-l \bmod N} \tag{A.3}$$

where $\tilde{h}_{j,l} = h_{j,l}/2^{(j/2)}$ are the MODWT wavelet filters and $\tilde{g}_{j,l} = g_{j,l}/2^{(j/2)}$ are the MODWT scaling filters. Hence the additive decomposition or the multi-resolution analysis (MRA) is expressed by,

$$X = \sum_{j=1}^{J_0} \tilde{D}_j + \tilde{S}_{J_0} \tag{A.4}$$

where,

$$\tilde{D}_{j,t} = \sum_{l=0}^{N-1} \tilde{h}_{j,l} \tilde{W}_{j,t+l \bmod N} \tag{A.5}$$

$$\tilde{S}_{j,t} = \sum_{l=0}^{N-1} \tilde{g}_{j,l} \tilde{V}_{j,t+l \bmod N} \tag{A.6}$$

A set of coefficients D_j with N samples can be obtained at each scale j. These coefficients capture the local fluctuations of X at each scale. S_{J_0} captures the overall "trend" of X. Adding D_j to S_{J_0}, for $j = 1, 2, \ldots, J_0$, gives an increasingly more accurate approximation of the now decomposed X. This additive reconstruction allows to predict each wavelet sub-series (D_j, S_{J_0}) separately and add the individual predictions to generate an aggregate prediction.

In time-series prediction, the MODWT should be performed incrementally where a wavelet coefficient at a position n is calculated from the samples at positions less than or equal to n, but never larger. If this is not adhered it is a form of peeking. The À Trous filtering scheme[2] briefly described below is used to do this. Consider a signal $X(1), X(2), \ldots, X(n)$, where n is the present time point and perform the following steps for a 5 level decomposition:

1. For index k sufficiently large, carry out the MODWT transform (A.3), (A.4), (A.5) and (A.6) on $X(1), X(2), \ldots, X(n)$.
2. Retain the coefficient values as well as the smooth values for the kth time point only: $D_1(k), D_2(k), \ldots, S_5(k)$. The summation of these values gives $X(k)$.
3. If $k < n$, set $k = k + 1$ and go to Step 1. This process produces an additive decomposition of the signal $X(k), X(k + 1), \ldots, X(n)$, which is similar to the À Trous wavelet transform decomposition on $X(1), X(2), \ldots, X(n)$.

---

[2] Mark J Shensa. The discrete wavelet transform: wedding the a trous and mallat algorithms. Signal Processing, *IEEE Transactions on*, 40(10):2464–2482, 1992.

Appendix B
Production Rules for Some Standard Technical Indicators

This appendix presents the feature generation flow for the technical indicators R, Aroon, RSI and Chaikin volatility respectively using grammar families 3 to 7 (Tables B.1, B.2, B.3, B.4 and B.5). In Chap. 4, it was explained that features were generated with 3 values $n = 6, 12, 24$ for F^+, F^-, H^+, L^-, i^+ and i^- (Tables B.4 and B.5).

Figure B.1 shows how the indicator R is generated using grammar family 3 in Table B.1. Aroon type technical indicators can be easily generated as follows. Although the original Aroon up indicator is $(N - i^+)/N$ it is understood that the information is captured by simple division i^+/N, hence the subtraction from 1 is ignored.

Invoking rule (1.b) : $<L5> ::= (<L3>) \div N$
Invoking rule (3.f) : $<L3> ::= <L1>$
Invoking rule (5.a) : $(<L1>) ::= i^+$
Invoking rule (5.b) : $(<L1>) ::= i^-$

Figure B.2 shows how the indicator RSI is generated using grammar family 5 in Table B.3. By defining $RS = ema(U, n)/ema(D, n)$, $RSI = RS/(1 + RS)$. By expanding this $RSI = ema(U, n)/(ema(U, n) + ema(D, n))$.

Table B.1 Grammar family 3

Family 3		
$\mathcal{N} = \{L1, L2, L3\}$		
$\mathcal{T} = \{-, \div, \text{lag}, \text{sma}, \text{meandev}, \text{sum}, H_h, L_l, C, n, k, (\,,\,)\}$		
$\mathcal{S} = \{L3\}$		
Production rules: $\mathcal{R}$		
$\langle L3 \rangle$	$::= ((\langle L2 \rangle)) \div ((\langle L2 \rangle)) \mid \text{sma}(\langle L2 \rangle, n) \mid \langle L2 \rangle$	(1.a), (1.b), (1.c)
$\langle L2 \rangle$	$::= \langle L1 \rangle - \text{lag}(\langle L1 \rangle, k) \mid \text{sma}(\langle L1 \rangle, n)$	(2.a), (2.b)
	$\mid \text{meandev}(\langle L1 \rangle, n) \mid \text{sum}(\langle L1 \rangle, n) \mid \langle L1 \rangle$	(2.c), (2.d), (2.e)
$\langle L1 \rangle$	$::= H^+ \mid L^- \mid C$	(3.a), (3.b), (3.c)

© The Author(s) 2015
A.M. De Silva and P.H.W. Leong, *Grammar-Based Feature Generation for Time-Series Prediction*, SpringerBriefs in Computational Intelligence, DOI 10.1007/978-981-287-411-5

Table B.2 Grammar family 4

Family 4

$\mathcal{N} = \{L1, L2, L3, L4, L5\}$

$\mathcal{T} = \{-, \div, \text{lag}, \text{ema}, \text{sma}, \text{meandev}, \text{sum}, H, L, C, n, k, i^+, i^-, (,)\}$

$\mathcal{S} = \{L5\}$

Production rules: $\mathcal{R}$

$\langle L5 \rangle$	$::= (\langle L3 \rangle \div \langle L4 \rangle) \,	\, (\langle L3 \rangle \div N) \,	\, \langle L4 \rangle$	(1.a), (1.b), (1.c)
$\langle L4 \rangle$	$::= \text{ema}(\langle L1 \rangle, n) \,	\, \text{sum}(\langle L1 \rangle, n) \,	\, \max(\langle L1 \rangle, n)$	(1.d), (1.e)
	$\,	\, \min(\langle L1 \rangle, n) \,	\, (\langle L1 \rangle) \div N \,	\, \langle L1 \rangle$	(2.a), 2.b), (2.c)
$\langle L3 \rangle$	$::= \langle L2 \rangle - \text{ema}(\langle L2 \rangle, n) \,	\, \text{ema}(\langle L2 \rangle, n) \,	\, \text{meandev}(\langle L2 \rangle, n)$	(3.a), (3.b)
	$\,|\, \text{sum}(\langle L2 \rangle, n) \,|\, \max(\langle L2 \rangle, n) \,|\, \min(\langle L2 \rangle, n) \,|\, \langle L1 \rangle$	(3.c), (3.d), (3.e), (3.f)
$\langle L2 \rangle$	$::= H \,	\, L \,	\, C$	(4.a), (4.b), (4.c)
$\langle L1 \rangle$	$::= i^+ \,	\, i^-$	(5.a), (5.b)

Table B.3 Grammar family 5

Family 5

$\mathcal{N} = \{L1, L2, L3\}$

$\mathcal{T} = \{-, \div, \text{lag}, \text{ema}, \text{sma}, \text{meandev}, \text{sum}, H, L, C, n, k, i^+, i^-, (,)\}$

$\mathcal{S} = \{\text{expr}\}$

Production rules: $\mathcal{R}$

$\langle L5 \rangle$	$::= \langle L3 \rangle \div (\langle L3 \rangle + \langle L4 \rangle) \,	\, \langle L3 \rangle \div (\langle L3 \rangle - \langle L4 \rangle)$	(1.a), (1.b), (1.c)
$\langle L4 \rangle$	$::= \text{ema}(\langle L1 \rangle, n) \,|\, \text{sum}(\langle L1 \rangle, n) \,|\, \text{meandev}(\langle L1 \rangle, n) \,|\, \max(\langle L1 \rangle, n)$	(2.a), (2.b)
	$\,|\, \min(\langle L1 \rangle, n) \,|\, \text{delt}(\langle L1 \rangle)$	(2.c), (2.d), (2.e)
$\langle L3 \rangle$	$::= \text{ema}(\langle L2 \rangle, n) \,|\, \text{sum}(\langle L2 \rangle, n) \,|\, \text{meandev}(\langle L2 \rangle, n) \,|\, \max(\langle L2 \rangle, n)$	(3.a), (3.b)
	$\,|\, \min(\langle L2 \rangle, n) \,|\, \text{delt}(\langle L2 \rangle)$	(3.c), (3.d), (3.e)
$\langle L2 \rangle$	$::= F^- \,	\, D$	(4.a), (4.b)
$\langle L1 \rangle$	$::= F^+ \,	\, U$	(5.a), (5.b)

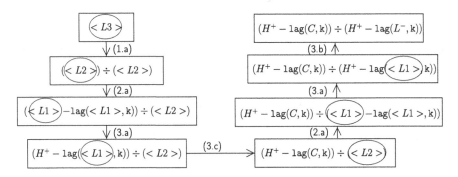

Fig. B.1 Feature generation flow for the technical indicator R

Table B.4 Grammar family 6

Family 6

$\mathcal{N} = \{$L1, L2, L3$\}$

$\mathcal{T} = \{-, \div, \text{lag}, \text{ema}, \text{sma}, \text{meandev}, \text{sum}, \text{H}, \text{L}, \text{C}, \text{n}, \text{k}, i^+, i^-, (,)\}$

$\mathcal{S} = \{$L4$\}$

Production rules: $\mathcal{R}$

$\langle L4 \rangle$	$::= ((\langle L1 \rangle - \langle L3 \rangle) \div ((\langle L2 \rangle - \langle L3 \rangle) \mid \langle L2 \rangle \mid \langle L3 \rangle$	(1.a), (1.b), (1.c)
$\langle L3 \rangle$	$::= \text{sma}(\langle L1 \rangle, \text{n}) - 2 \times \text{sd}(\langle L1 \rangle, \text{n})$	(2.a)
$\langle L2 \rangle$	$::= \text{sma}(\langle L1 \rangle, \text{n}) + 2 \times \text{sd}(\langle L1 \rangle, \text{n})$	(3.a)
$\langle L1 \rangle$	$::= \text{H} \mid \text{L} \mid \text{C} \mid \text{H-L} \mid \text{H-C} \mid \text{C-L}$	(4.a), (4.b), (4.c), (4.d), (4.e), (4.f)

Table B.5 Grammar family 7

Family 7

$\mathcal{N} = \{$L1, L2, L3$\}$

$\mathcal{T} = \{-, \div, \text{lag}, \text{sma}, \text{H}, \text{L}, \text{C}, \text{M}, (,)\}$

$\mathcal{S} = \{$expr$\}$

Production rules: $\mathcal{R}$

$\langle L3 \rangle$	$::= ((\langle L2 \rangle) \div ((\langle L2 \rangle) \mid ((\langle L2 \rangle - \langle L2 \rangle) \mid \langle L2 \rangle$	(1.a), (1.b), (1.c)
$\langle L2 \rangle$	$::= \text{ema}(\langle L1 \rangle, \text{n}) \mid \text{sma}(\langle L1 \rangle, \text{n}) \mid \text{wma}(\langle L1 \rangle, \text{n})$	(2.a), (2.b), (2.c)
	$\mid \text{sma}(\text{ema}(\langle L1 \rangle, \text{n}), \text{n}) \mid \langle L1 \rangle$	(2.d), (2.e)
$\langle L1 \rangle$	$::= \text{lag}(\text{V}, \text{k})$	(3.a)

Fig. B.2 Feature generation flow for the RSI indicator

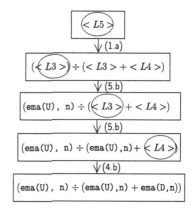

The upper band of Chaikin volatility is generated as follows using grammar family 6 in Table B.4.

Invoking rule (1.c) : $<L4> ::= (<L3>)$
Invoking rule (2.a) : $<L3> ::= \text{sma}(<L1>, \text{n}) + 2 \times \text{sd}(<L1>, \text{n})$
Invoking rule (4.d) : $<L1> ::= \text{H-L}$

The lower band can be generated similarly by using rule (2.b) instead of (2.a).

Appendix C
Supplementary Results

C.1 Model-based Approaches on Stock Indices

Autoregressive integrated moving average (ARIMA) and exponential time-series smoothing (ETS) were used as pure time-series model benchmarks. `auto.arima()`, `Arima()` and `ets()` functions in the `forecast` package in R were used for implementation. ARIMA model order was automatically decided using `auto.arima()` for day-ahead predictions in the validation dataset. The error type, trend type and seasonal type of ETS models were automatically decided using `ets()`. Table C.1 gives the ARIMA model orders for the 10 indices.

C.2 SVM Parameter Values

The C and γ parameter of the SVM were chosen using a parallelized grid search. The grid range is specified in Table 6.2. Table C.2 gives the optimal parameters selected for SVM using technical indicators as features.

C.3 Foreign Exchange Client Net Trade Volume Binary Classification

Evaluation Criteria: For unbalanced data sets, accuracy alone is not a good criterion for evaluating a model performance. Therefore a range of measures were evaluated. Define TP $\equiv$ True Positive, TN $\equiv$ True Negative, FP $\equiv$ False Positive and TN $\equiv$ True Negative. Accuracy $\equiv$ Hit Ratio (HR) $=$ (TP $+$ TN)/(TP $+$ FP $+$ TN $+$ FN).

© The Author(s) 2015
A.M. De Silva and P.H.W. Leong, *Grammar-Based Feature Generation for Time-Series Prediction*, SpringerBriefs in Computational Intelligence, DOI 10.1007/978-981-287-411-5

Table C.1 ARIMA model order for the 10 indices

GSPC	1,1,1	SSMI	4,1,5	N225	2,1,0	GDAXI	4,1,4	HSI	2,1,2
FTSE	4,1,4	SSEC	0,1,0	NDX	4,1,4	TWII	1,1,0	AORD	0,1,0

Table C.2 Parameters for SVM using TIs as features with $\varepsilon = 0.01$

Symbol	C	γ
AORD	190	$5.823249e - 05$
FTSE	70	0.00049476
GDAXI	200	$8.99928e - 05$
GSPC	110	0.00014141
HSI	200	0.0001979628
TWII	140	0.00049476
NDX	190	0.0002474412
N225	190	0.000109989
SSEC	190	$7.614881e - 05$
SSMI	170	$8.249402e - 05$

Sensitivity = TP/(TP + FN), Specificity = TN/(TN + FP) and Balanced Accuracy (BAC) = (Sensitivity + Specificity)/2. Precision = TP/(TP + FP), Recall = TP/(TP + FN) and F-score = 2 × Precision × Recall/(Precision + Recall) (Table C.3).

Table C.3 Out-sample results (%) using integer GA to predict the client trade volume direction for 6 months using SVM

Evaluation criterion	Month 1	Month 2	Month 3	Month 4	Month 5	Month 6
Confusion matrix	210 30	206 26	235 13	261 8	234 5	225 7
	97 15	87 17	97 7	65 2	96 1	109 11
Sensitivity	13.39	16.34	6.73	2.98	1.03	9.16
Specificity	87.50	88.79	94.76	97.03	97.91	96.98
Balanced accuracy	50.44	52.57	50.74	50.01	49.47	53.07
Precision	33.33	39.53	35.00	20.00	16.67	61.11
Recall	13.39	16.35	6.73	2.98	1.03	9.17
F-Score	19.11	23.13	11.29	5.19	1.94	15.94
Hit ratio	63.92	66.37	68.75	78.27	69.94	67.04

C.4 Dominant Features Providing Best Results in the SVM for Chosen Indices

See Tables C.4 and C.5.

Table C.4 Feature frequency for GSPC (*left*) and SSMI (*right*) using wrapper based GA for feature selection

Feature	Parameters	Frequency
$\sigma(\Delta H_k, 15)/\sigma(\delta L_k, 15)$		10
Disparity	$n = 5$	8
$(M_k - \Delta H_k)/15$		7
$SMA(L_k, n) + 2 \times \sigma(L_k, 30)$	$n = 5, 15, 30$	6
$(C_k - (M_k - H_{k-2}))/n$	$n = 5, 15$	5
$(H_k - (L_k - H_{k-2}))/n$	$n = 5, 15$	5
$C_k - SMA(\Delta H_k, 30)$		5
$H_k - SMA(EMA(\Delta C_k, 30))$		5
$SMA(C_k, 15) + 2 \times \sigma(C_k, n)$	$n = 5, 15, 30$	5
$SMA(H_k - \Delta C_k, n)$	$n = 15, 30$	5
$SMA(H_k, n) + 2 \times \sigma(H_k, 15)$	$n = 5, 15, 30$	5
$H_k - \sigma(M_{k-i}, 5)$	$i = 0, 2, 5, 6$	4
$H_{k-1} - SMA(\Delta H, 30)$		4
$L_k - SMA(EMA(\Delta L, n_1), n_2)$	$n_1 = 15, 30, n_2 = 5, 15, 30$	4
$\Sigma(L_k, n)/\max(i^+, 30)$	$n = 5, 15$	4
$C_k - SMA(EMA(\Delta L, n_1), n_2)$	$n_1 = 5, 15, n_2 = 15, 30$	4
H_{k-1}		3
$H_k - WMA(\Delta H_k, n_1)$	$n_1 = 5, 15$	3
$(L_k - (H_k - C_{k-i}))/n$	$n = 5, 15, i = 1, 2$	3
$(M_k - (L_k - C_{k-2}))/n$	$n = 5, 15$	3
$(C_k - (M_k - L_{k-i}))/n$	$n = 5, 15, i = 1, 2, 3, 5$	6
$(C_k - \bar{\sigma}(H_k, n_1))/n_2$	$n_1 = 5, 15, 30, n_2 = 5, 15$	6
$(C_k - (L_k - C_{k-i}))/n$	$n = 5, 15, i = 0, 1, 2, 3$	5
$(C_k - \sigma(M_{k-3}, n))$	$n = 5$	5
$\sigma(\Delta C_k, n_1)/\sigma(\delta M_k, n_2)$	$n_1 = 5, 15, 30, n_2 = 5, 15, 30$	5
$(C_k - (H_k - H_{k-i}))/n$	$n = 5, 15, i = 2, 3$	4
$C_k - \sigma(C_{k-i}, 5)$	$i = 3, 4, 5$	4
$\sigma(\Delta C_k, n_1)/\sigma(\delta C_k, n_2)$	$n_1 = 5, 15, n_2 = 5, 15$	4
$\Sigma(H_h(6), 5)/C_k$		4
$(C_k - (H_k - L_{k-i}))/n$	$n = 5, 15, i = 1, 3$	3
$(C_k - (H_k - SMA(L_k, n_1)))/n_2$	$n_1 = 5, 30, n_2 = 5, 15$	3
$(C_k - (L_k - H_{k-2}))/5$		3

(continued)

Table C.4 (continued)

Feature	Parameters	Frequency
$(C_k - (L_k - L_{k-i}))/n$	$n = 5, 15, i = 1, 3$	3
$(C_k - (L_k - M_{k-i}))/5$	$i = 2, 3$	3
$(C_k - (M_k - L_{k-2}))/15$		3
$(C_k - (M_k - \text{SMA}(H_k, n_1)))/n_2$	$n_1 = 5, 15, n_2 = 5, 15$	3
$(C_k - \tilde{\sigma}(L_k, 30))/15$		3
$(C_k - \tilde{\sigma}(M_k, n_1))/n_2$	$n_1 = 5, 30, n_2 = 5, 15$	3
$(H_k - (L_k, H_{k-i}))/15$	$i = 0, 2, 3$	3
$(H_k - \tilde{\sigma}(C_k, n))/5$	$n = 5, 30$	3

Table C.5 Feature frequency for FTSE (*left*) and HSI (*right*) when using wrapper based GA for feature selection

Feature	Parameters	Frequency
$L_k - \sigma(H_{k-i}, n)$	$n = 5, 15, i = 1, 2, 4, 5$	6
$L_k - \sigma(M_{k-i}, n)$	$n = 5, 15, i = 0, 1, 3, 5$	6
$C_k - \sigma(L_{k-i}, 5)$	$i = 3, 5$	5
$C_k - \sigma(M_{k-i}, n)$	$n = 5, 15, i = 0, 2, 3, 6$	5
$M_k - \sigma(C_{k-i}, n)$	$n = 5, 15, i = 0, 1, 3, 6$	5
$C_k - (L_k - \text{SMA}(L_k, 30))/5$		4
$C_{k-i} - \Delta H_k$	$i = 0, 1, 2$	4
$C_k - \sigma(C_{k-i}, n)$	$n = 5, 15, i = 0, 3, 5$	4
$L_k - \sigma(C_{k-i}, n)$	$n = 5, 15, i = 3, 5, 6$	4
$M_k - \text{SMA}(\text{EMA}(\Delta L, n_1), n_2)$	$n_1 = 5, 15, n_2 = 5, 15, 30$	4
$(M_k - (L_k - H_{k-i}))/n$	$n = 5, 15, i = 1, 2, 3$	4
Bias		3
$(C_k - (H_k - H_{k-i}))/n$	$n = 5, 15, i = 1, 3$	3
$(C_k - (L_k - C_{k-i}))/15$	$i = 0, 2, 3$	3
$(C_k - \tilde{\sigma}(C_{k-i}, n_1))/5$	$n_1 = 15, 30$	3
Disparity		3
$C_{k-i} - \sigma(\Delta H, 5)$	$i = 2, 3$	3
$C_k - \sigma(H_{k-i}, n)$	$n = 5, 15, i = 1, 3, 6$	3
$M_k - \sigma(L_{k-i}, 5)$	$i = 3, 4, 5$	3
$(L_k - (H_k - \text{SMA}(L_k, 5)))/n$	$n = 5, 15$	3
$C_k - \sigma(M_{k-i}, n)$	$n = 5, 15, i = 2, 6$	4
$L_k - \sigma(H_{k-i}, n)$	$n = 5, 15, i = 1, 3, 6$	4
$(L_k - (H_k - L_{k-i}))/n$	$n = 5, 15, i = 0, 2, 3$	4
ROC		4
$\text{SMA}(M_k, n_1) + 2 \times \sigma(M_k, n_2)$	$n_1 = 5, 15, n_2 = 5, 15, 30$	4

(continued)

Table C.5 (continued)

Feature	Parameters	Frequency
Aroon		3
$(C_k - (M_k - H_{k-1}))/n$	$n = 5, 15$	3
$(C_k - \bar{\sigma}(H_k, 5))/n$	$n = 5, 15$	3
Disparity		3
$(H_k - \bar{\sigma}(L_k, n))/15$	$n = 5, 15$	3
$C_k - \sigma(\Delta M_k, 5)$		3
$C_k - \text{SMA}(\Delta C_k, n)$	$n = 5, 30$	3
$C_k - \text{SMA}(\Delta H_k, n)$	$n = 5, 30$	3
$H_k - \sigma(\Delta L_k, n)$	$n = 5, 15, 30$	3
L_{k-6}		3
$L_k - \sigma(L_{k-i}, 5)$	$i = 1, 2, 6$	3
$L_k - \sigma(L_{k-i}, n)$	$n = 5, 15, i = 1, 5, 6$	3
$M_k - \sigma(H_{k-i}, n)$	$n = 5, 15, i = 1$	3
$M_k - \sigma(L_{k-i}, n)$	$n = 5, 15, i = 1, 2, 6$	3
$M_k - \sigma(M_{k-i}, n)$	$n = 5, 15, i = 1, 2$	3

Printed in the United States
By Bookmasters